THE MICROSTRUCTURE OF DINOSAUR BONE

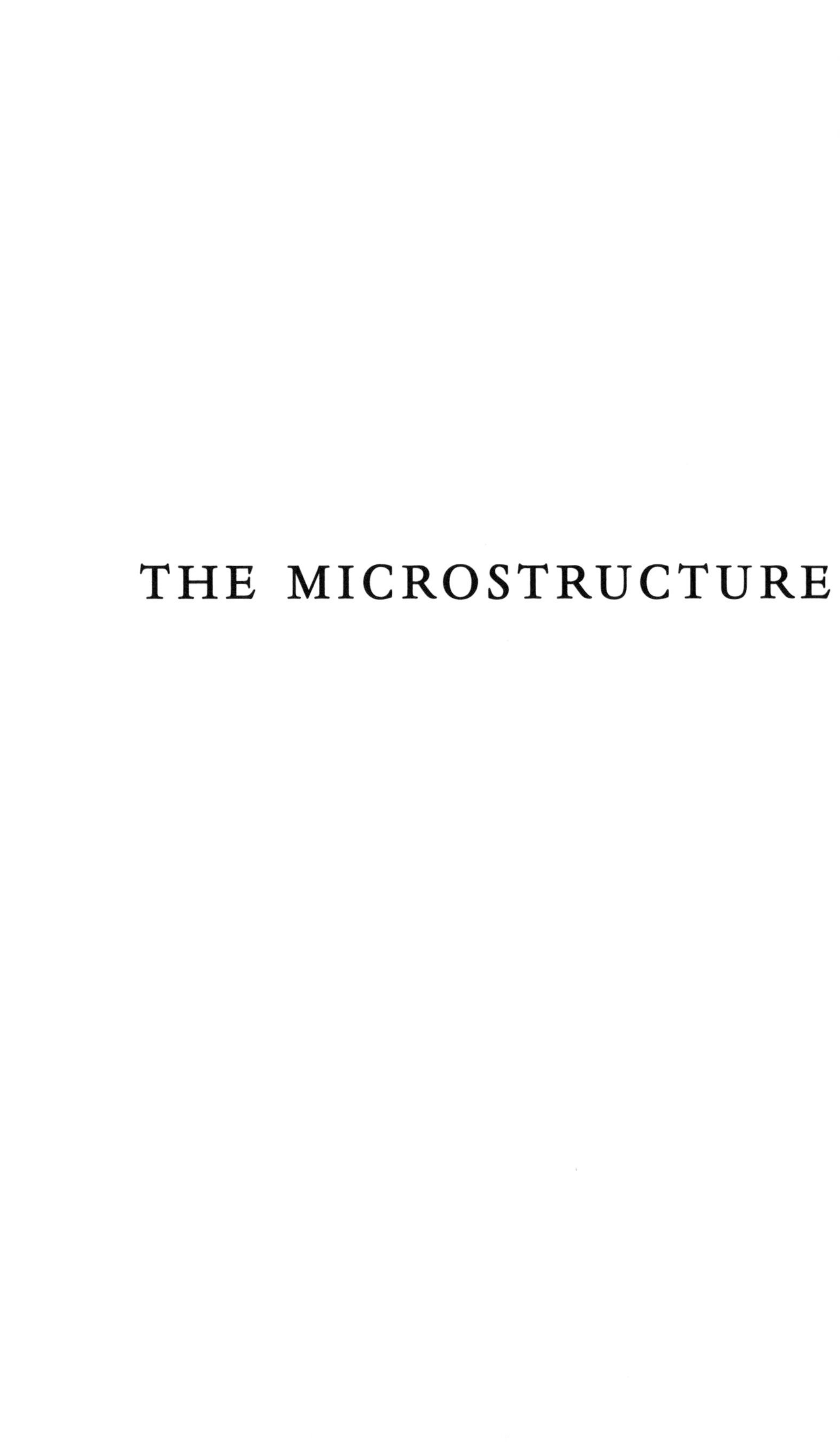

THE MICROSTRUCTURE

OF DINOSAUR BONE

Deciphering Biology with Fine-Scale Techniques

Anusuya Chinsamy-Turan

The Johns Hopkins University Press
Baltimore and London

Printed in the United States of America on acid-free paper
9 8 7 6 5 4 3 2 1

The Johns Hopkins University Press
2715 North Charles Street
Baltimore, Maryland 21218-4363
www.press.jhu.edu

Library of Congress Cataloging-in-Publication Data
Chinsamy-Turan, Anusuya.
The microstructure of dinosaur bone : deciphering biology with fine-scale
techniques / Anusuya Chinsamy-Turan.
p. cm.
Includes bibliographical references and index.
ISBN 0-8018-8120-X (hardcover : alk. paper)
1. Reptiles, Fossil. 2. Dinosaurs. 3. Dinosaurs—Research. I. Title.
QE861.4.C48 2005
567.9—dc22 2004023000

A catalog record for this book is available from the British Library.

For Yunus, Evren, and Altay

Contents

Preface

The microscopic structure of dinosaur bone survives millions of years of fossilization, and in its nature and texture, it holds an unparalleled record of growth processes the animal experienced during life. A synthesis of more than 150 years of research is provided in this first-ever book on dinosaur bone microstructure, and the effect of such studies on the reconstruction of dinosaurs as once-living animals is realized. This book also explores the microscopic structure of bones of the closest living relatives of dinosaurs (i.e., birds and crocodiles) and a range of other extant and extinct vertebrates. An "insider's" perspective of how the field has grown and possibilities for new research are highlighted.

Chapter 1 deals with the bone structure and composition in a living dinosaur and how the bones change once the animal dies. What enables bone's resilience in the fossil record is considered.

In chapter 2, phylogenetic, paleoenvironmental, and biological interpretations of dinosaurs' skeletal remains are addressed. The extant phylogenetic bracket approach that enables reliable deductions about soft tissue and behavior in fossil taxa is expounded on. In addition, this chapter considers various macroscopic, microscopic, and molecular studies that can be conducted on dinosaur bones and provides detailed guidelines regarding selection of samples for microstructural analyses and outlines procedures for thin-sectioning bones.

Chapter 3 delves into the nature of the microscopic structure of extant vertebrate bones, focuses on the factors that affect bone deposition in modern animals, and considers whether signatures of such effects are recorded in bone and, if so, considers how they can be used to infer various aspects of the biology of dinosaurs and other extinct animals.

Chapter 4 documents bone tissue types that are readily recognizable in dinosaurs, and their usefulness in deciphering dinosaur growth patterns and biology. A historical perspective and a synthesis of studies in dinosaur bone microstructure are provided.

Chapter 5 closely inspects how the fossilized bones of dinosaurs provide information about how dinosaurs grew from embryonic stages to adulthood.

Chapter 6 provides a succinct overview of Mesozoic birds and examines what is known about early bird biology. Phylogenetic and ontogenetic studies of the bone microstructure of basal birds are explored, and their effect on understanding basal bird biology and growth are considered.

The final chapter investigates the evidence for interpreting dinosaur physiology, and the perennial question of endothermic (hot-blooded) versus ectothermic (cold-blooded) dinosaurs is addressed. How aspects of bone microstructure have fueled these deliberations are scrutinized, and contributions of new lines of evidence are discussed. This chapter looks closely at what bone microstructure can and cannot tell about the physiology of dinosaurs.

This book is written in an accessible manner for readers at various levels. It begins at an introductory level, assuming little or no knowledge of bone microstructure, and gradually becomes more complex, while supplying clear explanations. The book presents new ideas and fresh perspectives and will serve as a valuable reference for studies on dinosaur bone microstructure. Students and professionals in vertebrate paleontology who seek to understand the nature of fossil bone microstructure and its relevance in interpreting various aspects of the biology of extinct taxa will find this book immensely useful.

Acknowledgments

Writing a book is not easy—so I have learned! I have also realized that the companionship of wonderful family, friends, and colleagues makes everything and anything possible. I count myself as extremely privileged to have had such a bounty of first-rate people to help me along. Let me begin at the beginning:

First and foremost, I am indebted to my parents Krishna and Sushilla for giving me the confidence and freedom to pursue my ambitions and for enabling me to overcome the many barriers of being a black woman in South Africa in the apartheid days. Judy Maguire and Michael Raath, thank you for opening my eyes, mind, and heart to paleontology. I am eternally grateful to the late James Kitching for the support and encouragement he bestowed on me and for sharing his enthusiasm and love of fossils with me. Thanks to Armand De Ricqlès for helping me get started in bone microstructure studies and for guiding my early research. I am indebted to Robin Reid for scholarly advice and for the many stimulating discussions on bone microstructure. I thank Peter Dodson for his support and encouragement and for being such an exceptional friend and colleague.

Thanks to Ginger Berman (then of the Johns Hopkins University Press) for having put the idea of a "book" into my head and to Vincent Burke for having helped me finish it. I am indebted to Kholeka Mvumvu-Sidinile and Kerwin von Willingh for providing technical assistance and more recently to Natasha Kruger for getting me over the final technical hurdles. Sanghamitra Ray, who spent two years as a postdoctoral fellow in my laboratory, is acknowledged for drawing some of the diagrams. Thanks to Andrea Plos for always being just a phone call away and for rescuing me so many times from IT disasters.

I am indeed richer (and a whole lot wiser!) for having had so many good friends and colleagues who have supported and contributed toward my research endeavors. In this regard, I am thankful to Luis Chiappe, Sankar Chatterjee, Peter Cook, Phil Currie, Billy De Klerk, Andrzej Elzanowski, Cathy Forster, Hailu You, Jaap Hillenius, Zofia Kielan-Jaworowska, Jorn Hurum, Makoto Manabe, Larry Martin, John Ruben, Bruce Rubidge, Paul Sereno, Daya Reddy, Matthias Starck, David Unwin, Dong Zhiming, and Zhonghe Zhou. I am especially grateful to Peter Dodson and Sanghamitra Ray for having commented on an earlier draft of this book.

This book has benefited from the contributions of illustrations from several friends and colleagues, for which I am most grateful: Jennifer Botha, Luis Chiappe, Andrzej Elzanowski, Willem (Jaap) Hillenius, Zofia Kielan-Jaworowska, Gerhard Marx, Robin Reid, Martin Sander, Matthias Starck, Clive Truman, Allison Tumarkin-Deratzian, Larry Witmer, and Zhonghe Zhou.

Because I wrote this book with my two young boys keeping me on my toes and while holding a full-time job, it is an understatement to say, "I always needed time." I am therefore, extremely grateful to Mary Brown, for baby-sitting and giving me the much-needed time to work on this book. To my friends Pavs Pillay, Susan Rosendorff, Medee Rall, Razina Salie, and Himla Soodyall, thank you for standing beside me. I am especially grateful to my sisters, Saloshana and Siron, and their respective families for their enduring support in all my ventures.

Finally, without the encouragement, support, and understanding of my husband, Yunus Turan, this book would still be just a dream.

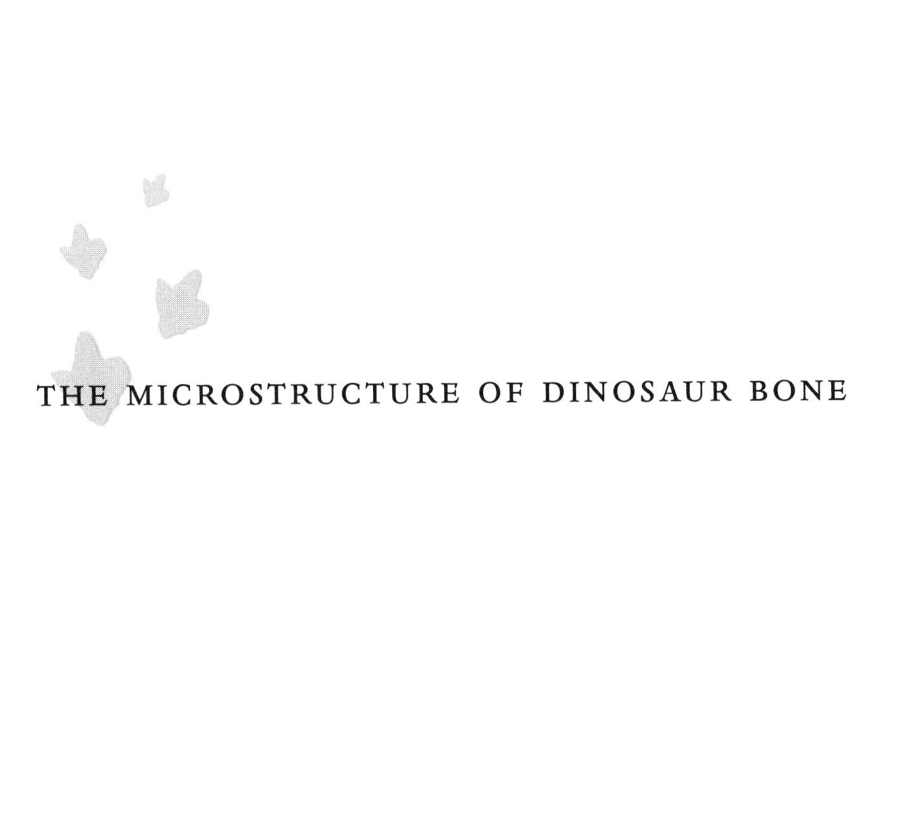

THE MICROSTRUCTURE OF DINOSAUR BONE

CHAPTER ONE

Unraveling Dinosaur Bones

Dinosaurs lived on earth for a phenomenal 160 million years. Yet today, aside from their modern descendants (birds), only the fossil record bears testimony to their spectacular diversity. Most of our knowledge of dinosaurs stems from the fossilized remains of bones and teeth. On occasion, dinosaur nests and tracks are found and rarely soft tissue is preserved (e.g., skin impressions, feathers, etc.). Though systematic studies of dinosaurs and paleobiogeographical analyses have led to a reasonable understanding of their evolution and phylogenetic relationships, there is still much that we do not know about dinosaurs as living animals.

In recent years, various aspects of dinosaur biology have been teased out of their fossilized remains through the burgeoning studies of dinosaur bone microstructure. It is astounding to realize that, after million of years of being subjected to various taphonomic processes, it is not just the morphology of dinosaur bones that is preserved but their microscopic structure as well.

Among living vertebrates, the microscopic structure of bones records the effect of various factors (such as the rate at which bone forms, ambient environmental conditions, and biomechanical and lifestyle adaptations) on their growth. Thus, by comparing dinosaur bone microstructure with that of extant vertebrates, and particularly with their extant relatives (i.e., birds and crocodiles), we have an unparalleled opportunity to deduce and understand the various factors that affected their growth.

Composition of Bone in a Living Dinosaur

The bones of all vertebrates have the same basic components. These are simply rearranged and reorganized depending on the type of vertebrate and on the rate at which the bone forms. Thus, at the molecular level, the basic composition of bone is the same in a dinosaur as in any other extinct or extant vertebrate. There are several histology textbooks that deal comprehensively with the microstructure of bone, so I examine it only briefly here. Instead, more emphasis is placed on the molecular organization of bone, which is rarely dealt with in conventional texts on bone. It is the organization of bone at this ultrastructural level that gives bones resilience in the fossil record.

Bone is a highly specialized connective tissue that constitutes the main calcified component of vertebrate skeletons. Soon after the discovery of the light microscope in the seventeenth century, knowledge of the structure of bone rapidly grew, and within a few decades, bone structures visible using ordinary light microscopy and measuring up to a few microns were described by various researchers. The steady growth of understanding bone structure parallels the technological advancements in microscopy (i.e., from light microscopy to polarized light microscopy and scanning and transmission electron microscopy). At each stage, current researchers used existing technologies to expand the understanding of the nature and structure of bone.

Bone is a composite material consisting of an organic and an inorganic phase. Bone-producing cells, called *osteoblasts,* deposit the organic phase (or *organic matrix* as it is often referred to) primarily composed of collagen. The collagen then "aggregates" in the extracellular space to form a framework onto which the inorganic mineral phase, a carbonated calcium phosphate ($Ca_{10}(PO_4CO_3)_6OH_2$)—a type of apatite—is deposited. The plate-shaped crystals of the bone mineral are about 500 Å long, 280 Å wide, and 20 Å thick and are regarded as the smallest known biologically formed crystals (e.g., Lowenstam and Weiner 1989; Fig. 1.1).

As early as 1936, Schmidt used polarized light microscopy to show that the *c* crystallographic axes of the mineral phase aligned with the collagen fibers, which pointed to a close association between the collagen and the mineral phase. With the advent of transmission electron microscopy, more details about the fine structure of bone emerged, and it became apparent that to understand the nature of bone it was essential to understand the nature of collagen and its association with the inorganic mineral phase of bone.

Organization of Collagen and Bone Mineral

Collagen in bone consists of three polypeptide chains (two of which are α_1 and the other α_2), dominated by 338 contiguous repeating triplet sequences in which every third amino

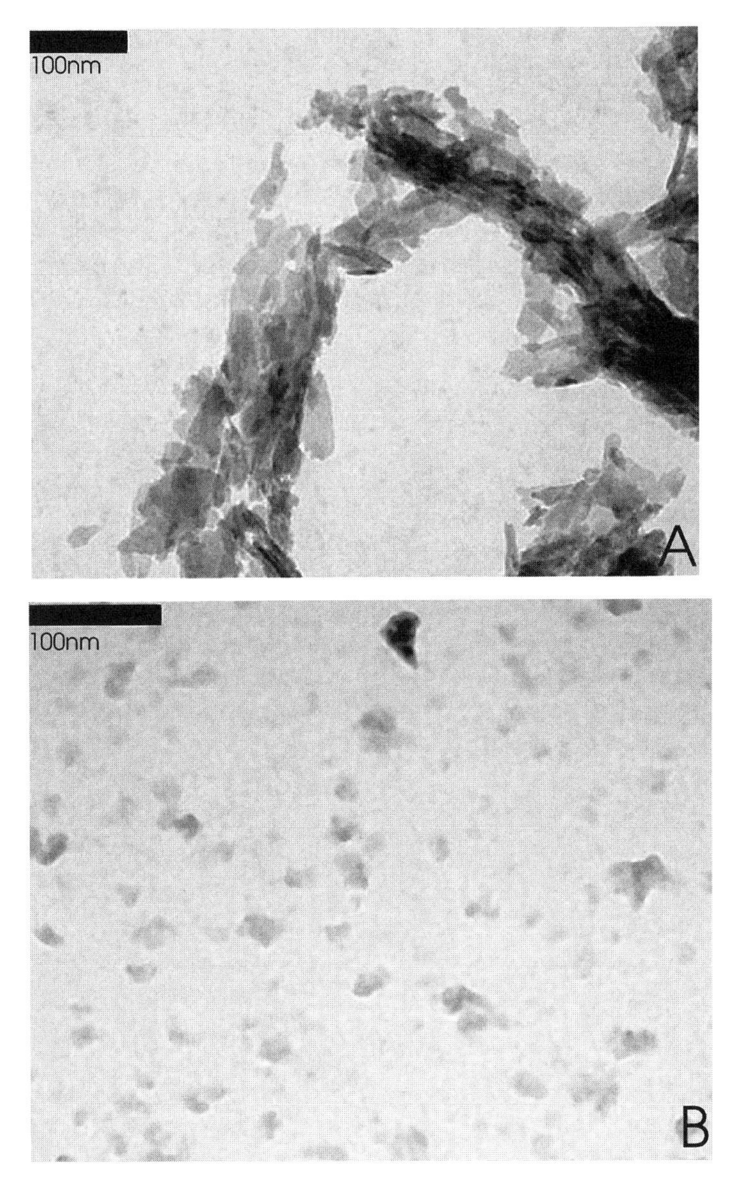

Fig. 1.1. *A*. Mineralized fibril from an elephant rib. *B*. Well-disaggregated crystals from a rib of a black rhino. Images courtesy of Clive Truman.

acid is a glycine. The triplet repeating regions of the three polypeptide chains fold together and result in a triple helical structure about 3000 Å long with a diameter of 15 Å, which forms the basic building block of the large collagen fibril and the framework on and within which the mineral crystals (i.e., the inorganic phase) are later deposited (Fig. 1.2).

Hodge and Petruska (1963) proposed the staggered array model for collagen that suggested that that the triple helical molecule "aggregate" into linear arrays separated by a hole or gap (Fig. 1.2). They proposed that the holes or gaps were the most likely sites for the mineral crystals. However, Lowenstam and Weiner (1989) point out that the holes are only 15–20 Å in diameter and some 360 Å long, which means that the mineral crystals could not only be located in the holes or gaps between adjacent polypeptide chains. According to Lowenstam and Weiner (1989), studies of bone using electron microscopy and x-ray diffraction analysis now suggest that a major portion of crystals are associated at the level of the gap or hole of the collagen fibrils and that, as the crystals grow, the overlap regions are also mineralized. Recent calculations of the available space for crystals suggest that 20%–30% of the mineral lies outside the fibers. It is now also known that the structure of collagen can change depending on the aqueous medium in which it occurs and on the prevalent osmotic pressure. Such changes can therefore affect the available space for the mineral crystals.

Although research is still continuing regarding the actual location of the crystals, it is now generally accepted that collagen forms the framework in which the apatite crystals align themselves with *c*-axis parallel to the collagen fibril long axis. This highly complex mineral-organic composite material constitutes the basic component of bone and becomes organized into layers, or lamellae, about 3–7 microns thick, which are arranged into higher-order structures, depending on the type of bone (Weiner and Traub 1992; Fig. 1.2). The intimate association of collagen and mineral gives bones their unique mechanical properties: the organic phase (collagen) provides flexibility to bone, while its hardness, strength, and rigidity are imparted by the inorganic mineral phase.

Organization at the Microstructural Level

In the living dinosaur within the bone matrix lie small spaces called *lacunae* in which osteocytes ("bone cells") occur (see Fig. 1.2). The osteocytes communicate with neighboring cells via cytoplasmic extensions that are housed in the canaliculi, which are extensions of the lacunae (Figs. 1.2, 1.3). The actual shape and size of the osteocytes vary, depending on the rate at which the bone is formed, and in addition, the number of canaliculi and the degree of branching are highly variable. The actual distribution (and orientation) of osteocytes depends on the rate at which the bone forms and, in the case of vascularized bones, on the type of vascularization that is present. For example, newly formed bone can simply incorporate blood vessels (forming simple blood vessels) or open spaces around blood vessels can be left, which later become infilled centripetally by more slowly forming bone tissue to form primary osteons. The actual

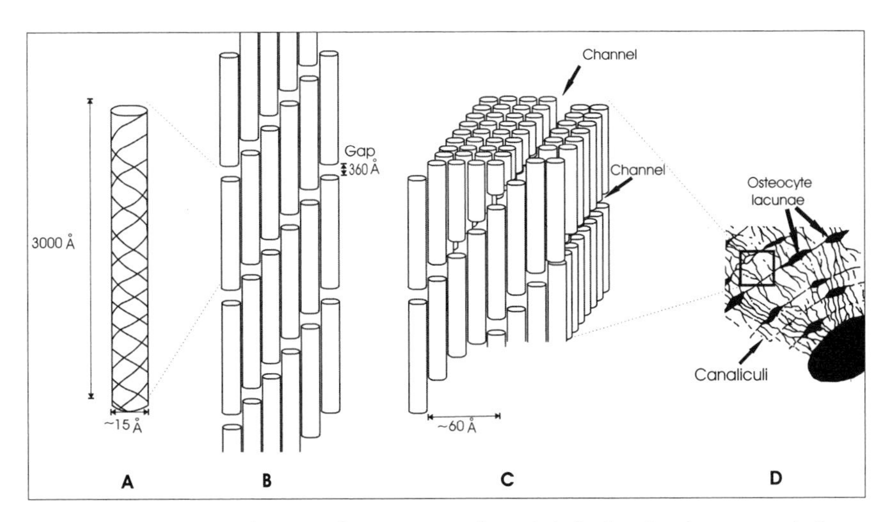

Fig. 1.2. The structure of type I collagen. *A.* A single triple helical molecule composed of two α_1 polypepetide chains and one α_2 chain. *B.* A two-dimensional section through part of a fibril showing the staggered array model. *C.* A three-dimensional perspective of the structure with the triple helical molecules arranged so that the holes form a channel or groove. *D.* The location of the collagen in the matrix of the bone. *A, B, C* are redrawn with modifications from Weiner and Traub (1992).

type of bone that forms and the amount of vascularization that occurs are influenced by a number of factors, including phylogeny and ontogeny.

On the basis of gross morphology, bone is classified as a dense bone called *compact bone* and a spongy or porous bone called *cancellous* bone. Cortical bone is recognized as the bone wall, or outer boundary or cortex of the bone, while cancellous bone is located within the core of the bone (Fig. 1.3). Depending on how bone develops, endochondral bone and intramembraneous bones are recognized. Endochondral bone develops from a cartilaginous precursor, while intramembraneous bones are formed directly from the mesenchyme. Three specialized types of intramembraneous bones are recognized: dermal bones that form directly through ossification of the mesenchyme; sesamoid bones that form within tendons; and perichondral, or periosteal, bone formed under the fibrous connective tissue covering cartilage (perichondrium) or bone (periosteum).

Upon Death of the Dinosaur

The taphonomic history of a dinosaur begins at the time of its death and continues until its discovery and, in most cases, even after discovery because the act of "collection"

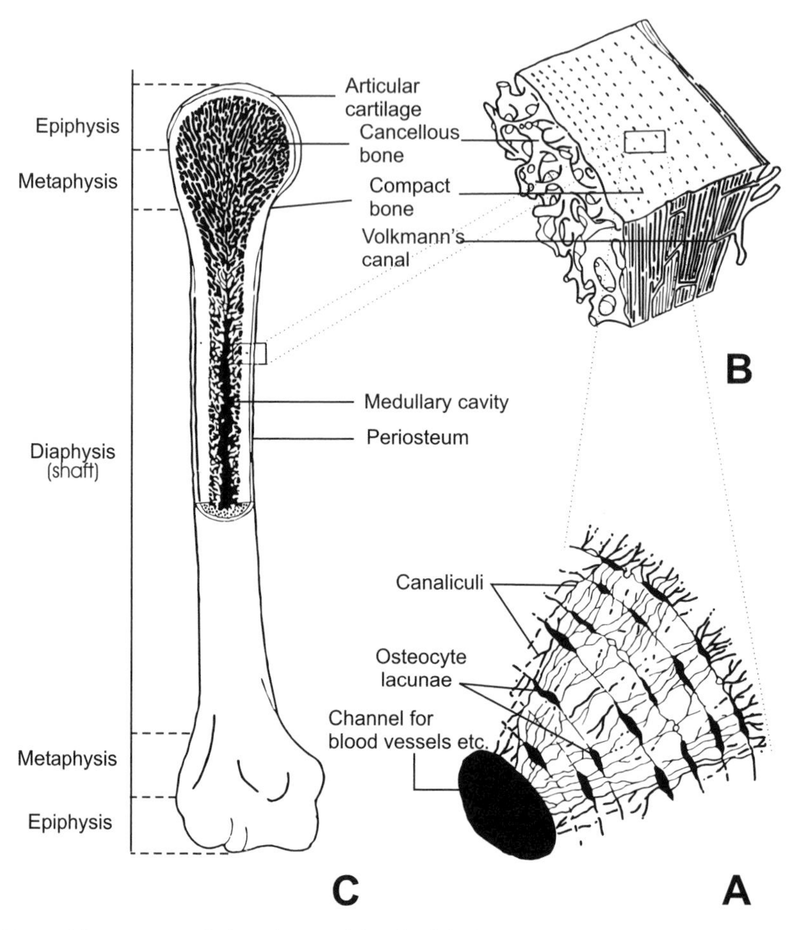

Fig. 1.3. The structure of a long bone and details of the bone microstructure (not drawn to scale). Fig. 1.2 showed the location of the collagen and mineral within the matrix of the bone, and *A* depicts this level of organization of bone. *B.* A higher level of organization. *C.* The gross overall morphology of the bone.

often affects the fossil. Taphonomy, therefore, considers all the natural processes of preservation and destruction of the remains of organisms, from death to burial and to its ultimate collection. The type of taphonomic processes that affect an animal is dependent on several factors that include its original morphology, behavior pattern, and life history, as well as its ecology, and how and where it actually died. Thus, taphonomy is an integral part of paleobiology, and any attempt to analyze a fossil assemblage has to consider the variety of taphonomic processes that affected the remains. A thorough analysis of taphonomic processes and effects is beyond the scope of this book, though there are several books that deal exclusively with this intriguing study (see, e.g., R. Lee Lyman's *Vertebrate Taphonomy* and Anna K. Behrensmeyer and Andrew Hill's *Fossils in the Making: Vertebrate Taphonomy and Paleoecology*).

Animals die for a number of reasons, as any modern ecological system will show. Natural causes, such as disease, old age, or predation, or even catastrophic mortality, such as a toxic volcanic gas emission, volcanic ash, flood, or mudslide, are just some of the causes of animal death. However, for a dinosaur to be preserved as a fossil a suite of conditions needs to be met. Most important, at death, the dinosaur has to have been buried quickly. Thus, the circumstances of death and the environment in which the animal dies are vitally important. A catastrophic mortality caused by cold weather will probably not involve rapid burial of the animal; however, a mudslide, volcanic ash fall, or a flood would enhance an animal's chance for burial and hence its preservation as a fossil. In addition, such fast burial may allow articulated remains to be preserved. Entrapment by quicksand, mud, and tarpits is another mode of death that may result in rapid burial, although in these cases, carcasses can be scavenged.

Let us assume for the sake of argument that a dinosaur was a victim of predation (i.e., it was hunted and killed). The predator would feed on the dinosaur prey; once the predator abandoned the carcass, scavengers would feed on it. Such carnivory causes dismemberment of the bones, disassociation, and dispersal of the remains of the animal. In addition, damage to the victim's bones occurs in the form of tooth marks, puncture wounds, and gnawing. Shortly after death, the dinosaur's soft tissues will begin to decompose. Decomposition usually proceeds from within the animal because of indigenous microflora, and then later proceeds from the outside once colonization of the carcass with microorganisms from the environment occurs. The decomposition of the soft tissue itself influences the immediate environment of the carcass, and it most likely affects the initiation and early manifestation of bone decomposition. Once skeletonization has occurred (i.e., soft tissue removed) the exposed skeleton generally becomes further disarticulated and is more easily dispersed. Bones are now subjected to insects and to natural weathering by sun, rain, wind, and acid. Thus, the sooner the remains of the dinosaur are covered and the risk of disarticulation and scattering are substantially reduced, the better its chances of entering the fossil record (Fig. 1.4).

The burial environment is crucial in the process of preservation and fossilization. Once buried, organic matter, including that within bones (such as collagen and other proteins, lipids, and connective tissue) and teeth, continues to decay. Eventually, only the inorganic parts of the animal, such as the crystalline, amorphous phase of bones

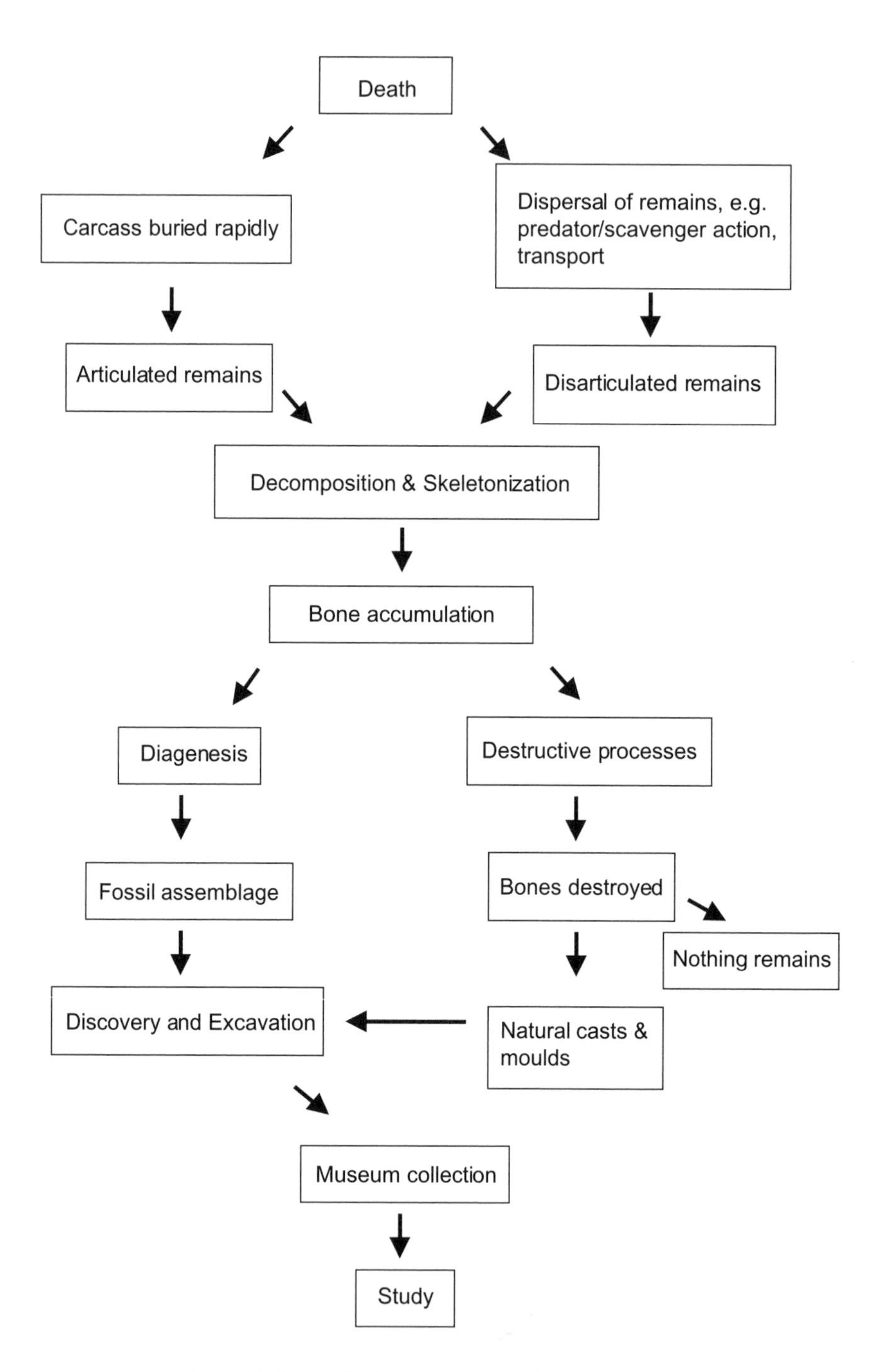

Fig. 1.4. Taphonomic history of an animal upon death.

and teeth, which are more resilient than the organic components, are left. The process of mineralization, termed *diagenesis* is extremely complex. Taphonomist Richard Lee Lyman of the University of Missouri–Columbia, describes the sum of the diagenetic processes (D) that a vertebrate skeleton undergoes as a function of M (the physical composition of the original material), C (the climate of decomposition), D_c (the decompositional mode), S (the nature of the sediment), and T (the time span), so that $D = f(M, C, D_c, S, T)$. Thus, these factors that affect diagenesis determine whether a fossil will eventually result.

On occasion, bones can be completely dissolved away by acidic solutions percolating through the rock. A different stone mineral can thereafter fill the vacant space and form a cast of the bones or material. It is however more common for the spaces to be unfilled and a mold of the animal results. This is the most common mode of preservation of *Mesosaurus,* an aquatic reptile from the Permian of Brazil and South Africa.

It should be obvious by now that the loss of a fossil can occur anytime during its taphonomic history. However, once buried in an environment that is conducive to preservation, the remains of the dinosaur generally stand a good chance of fossilization. After a while, compaction and cementation will result, which is characterized by a predominance of inorganic reactions. The skeletal remains of the entombed animal now have a strong chance of becoming a fossil, though during this stage of diagenesis, they can suffer compression, distortion, and fragmentation.

Changes in Bone Composition during Fossilization

It is common to find fossil bones deformed and distorted because of compression forces caused by the compaction and cementation phase of fossilization. However, sometimes fossil bones maintain the morphology of the living animal, so that the overall anatomy and bone structure, muscle scars, and pathologies are preserved. Either way, in all likelihood, the chemical composition of the bone has been altered during the fossilization process. Thus the macroscopic observation of a fossil bone is no indication of the nature of the histological and chemical preservation. The type of changes and the magnitude of the alterations that occur in the bones depend on the chemical factors in the environment, the properties of the sediment, and the composition of the actual skeleton.

In a skeleton, a bone is structured to perform its biomechanical function. This influences its mechanical and physical properties, which in turn affect the taphonomic processes that act on the bones. For example, the way in which bones fracture postmortem is directly related to the amount and distribution of the osteons and collagen fibers. If bones remain on the ground instead of being buried, subaerial weathering often causes cracks in the bones, which tend to be parallel to the long axis of the bone. On occasion, prolific mineral growth within medullary cavities of buried bones can cause cracks in the shaft of the bone, which also tend to form parallel to the long axis of the bone.

Most of the organic components of bone, such as collagen, decompose soon after burial; however, the inorganic fraction, mainly hydroxyapatite, which is a poorly crystalline calcium phosphate, is more resilient and is in general preserved with only slight modification. X-ray diffraction analyses of fossil bone often reveal that the crystalline, amorphous apatite part of hydroxyapatite is preserved, although the hydroxyl group is usually displaced by fluorine resulting in fluorapatite. Thus, the apatite part of the mineral is highly stable and undergoes relatively minor changes.

To understand the taphonomic history of ancient animals, modern analogues are essential. As such, experimental work on modern biota, considering factors that affect dispersal, scattering, transport, and bone modification, are important areas of taphonomic research. A leading researcher in this field is Anna Kay Behrensmeyer of the Smithsonian Institution in Washington, D.C. To document the macroscopic and microscopic changes that occur during the early stages of vertebrate taphonomic history, Behrensmeyer monitored marked carcasses in Amboseli Park, East Africa, over a period of about 10 years. After 10 years of surface exposure and natural weathering, samples of wildebeest bone were examined for proteins and alterations in the mineral component. Noreen Tuross from the Carnegie Geophysical Laboratory in Washington, D.C., performed the chemical analysis and found that the bones retained, albeit at diminished levels, original noncollagenous proteins and collagen alpha chains at their original molecular weights and recognizable epitopes. However, dinosaur bones are subjected to diagenetic alteration for substantially much longer and they rarely preserve collagen.

In modern bovine bone, when collagen is removed, the porosity of the bone increases by 26% (Trueman and Tuross 2003). Thus, given the close association of the collagen with the crystals of bone mineral during life, the dissolution of collagen in a fossil bone would theoretically expose the mineral, which, as expected, would lead to the dissolution of bone. However, as the fossil record shows, fossil bone is often preserved. Clive Trueman and Noreen Tuross found that well-preserved, unpermineralized dinosaur bone (i.e., bone with no minerals other than apatite) with no collagen preserved do not show the expected high-porosity levels. This suggests therefore that the space that was once occupied by collagen became secondarily infilled. Hubert (1996) has proposed that the space vacated by the collagen becomes infilled by apatite added to biogenic "seed" crystals, which maintains the apatite crystallite alignment (Fig. 1.5). This would explain the preservation in fos-

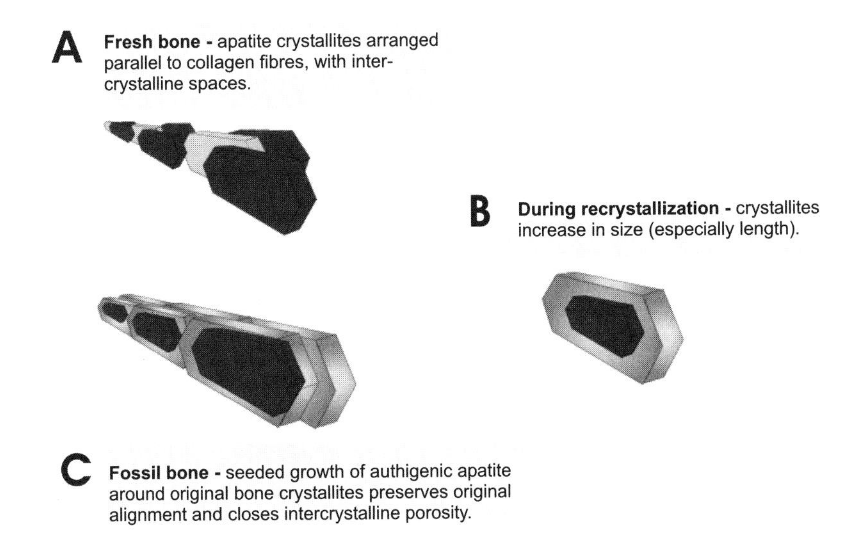

Fig. 1.5. The changes that the bone mineral undergoes during fossilization. Image courtesy of Clive Truman.

sil bones of the birefringence patterns observed in modern bone tissues. Crystal growth would continue until all the intercrystallite porosity is infilled by authigenic apatite—a process that would probably take about a thousand to hundreds of thousand of years (Trueman and Tuross 2003). Tuross found that after only 10 years the hydroxyapatite in the bones from Amboseli Park showed an increase in crystal size.

Although it is well recognized that bones undergo significant changes during fossilization, the exact nature of the changes are rather difficult to quantify. David Martill, (1991), a British taphonomist from the University of Portsmouth, recognizes three distinct areas wherein such changes are evident (i.e., within the bone tissue, within the pore spaces, and within the sediment around the bone). Another British researcher, Neil Garland (1987) from the University of Manchester, undertook a detailed paleohistological study of bone decomposition. Unlike Martill, he considers the bone tissue and pore spaces together and focuses on normal bone, destructive bone, and inclusions and infiltrations. For simplicity, Garland's categories are used below.

Normal Bone versus Destructive Changes in Bone

Thin sections of dinosaur bones often surprisingly reveal normal bone microstructure. By "normal," I mean a bone tissue that can be readily compared with the bone

Fig. 1.6. Diagenetically altered bone microstructure. *A*. A Triassic nonmammalian cynodont. *B*. *Morganucodon,* a Triassic mammal. *C*. An Early Cretaceous, hypsilophodontid polar dinosaur from Dinosaur Cove, Australia.

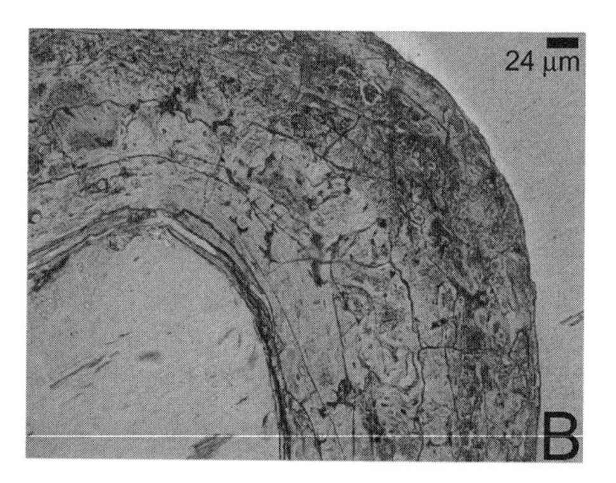

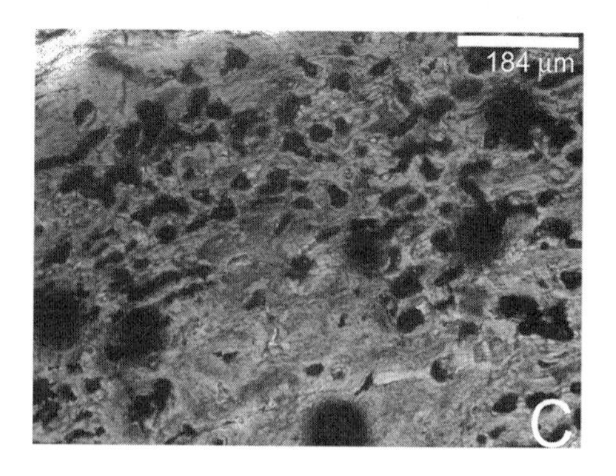

tissue of an extant vertebrate. The bone will show mineral infilling pore spaces, but the actual nature of the tissue is unaltered. Such "normal bone" is what is mainly used in paleohistological studies to assess various aspects of the biology of extinct animals. Quite often destructive changes in fossil bone are observable in thin sections as a disintegration, disaggregation, and dissociation of osteons. More focal loss of bone is described by Garland (1987) as rounded foci or interlinked tunnels. It appears that destructive changes in archaeological samples are fairly common; yet, in my experience, thin sections of paleontological aged fossil bones usually tend to preserve the microstructure and altered bone is rarely observed (Chinsamy 1997). This deduction has independently been supported by a recent paper by Clive Trueman and David Martill (2002) that suggests that, for bones to survive in the fossil record, bioerosion, which occurs early in postmortem and often leads to complete destruction of bone, needs to be prevented or halted. In instances in which fossil bones show some degradation of microstructure, it is quite likely that there was a change in the chemical conditions that served to inhibit the bioeroding microbes. In my thin-section collection of fossil bones, *Morganucodon,* an early mammal from the Triassic Pant fissures, shows the most extensive altered bone (Fig. 1.6). Although dinosaur bone microstructure is generally well preserved, on occasion examples of altered bone are found; for example, in my collection, I have examples of diagenetically altered bone from *Pachyrhinosaurus* from Alberta, from *Mononykus* from the Late Cretaceous of the Gobi Desert, and from a hypsilophondontid from the Early Cretaceous polar locality, Dinosaur Cove, in southeastern Australia (Fig. 1.6).

Inclusions in Fossil Bones

Inclusions are externally derived material found within the available bone spaces, such as the medullary cavity, vascular channels, osteocyte lacunae, and canaliculi. Inclusions can be biologically derived, such as algae, fungi, hyphae, bacteria, and insect parts or minerals. One of the most distinctive and extensive postdepositional damages to bone is that caused by microorganisms. Such changes to the bone cortex are readily observed under light microscopy and in microradiographic images. Alterations in the tissue structure include focal destruction and tunneling of the cortex, as well as mineral redeposition. Fungi of the genus *Mucor* is considered the primary destructive agent, although bacteria and algae have also been implicated in the postmortem destruction of bone.

In 1996, I and my colleagues (botanist Ed February, Eric Harley, who works on ancient DNA, and Australian paleontologists Tom Rich of the Victoria Museum and Pat

Vickers-Rich of Monash University) described the serendipitous inclusion of several pieces of wood within the marrow cavity of an Early Cretaceous femur of a hypsilophodontid from Dinosaur Cove, the "polar locality" in southeastern Australia (Fig. 1.7). We examined the wood under a scanning electron microscope and deduced that it belonged to *Podocarpus.* It is interesting that Tom Rich and Pat Vickers-Rich (1989) confirmed that *Podocarpus* was common in the Dinosaur Cove assemblage and therefore, the small bits of wood could easily have entered the marrow cavity of the dinosaur bone through a fracture in the bone wall.

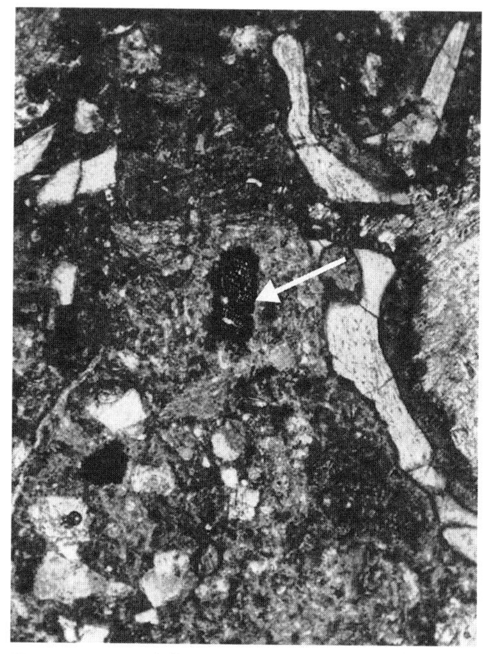

Fig. 1.7. A wood fragment (arrow) measuring only about 310 microns located inside the marrow cavity of a polar dinosaur bone.

It is however more common to find inorganic inclusions precipitating out of the ground water in the pore spaces of bones. Mineral inclusions are readily identified using transmitted or polarized light, as well as special stains, and commonly include calcite, pyrite, and quartz. In many instances, diagenetic minerals, or sediments, fill osteocyte lacunae and other pore spaces in the bone. Although the voids are secondarily infilled, the actual dimensions of the osteocyte lacunae, canaliculi, and vascular spaces remain relatively unaltered.

The type of mineral deposits that can occur within a fossil bone is dependent on the depositional environment and other factors that influence the degree of uptake of the minerals, for example, the size of the lumen to be filled or the orientation of the specimen, the thickness of the compact bone, the porosity of the bone wall, or simply whether the bone is fractured. In 1985, Daniel Brinkman and Michael Conway of Indiana University–Purdue University reported on a textural and mineralogical study that they had undertaken on a *Stegosaurus* plate. They found that the hydroxyapatite of the bone was altered to carbonate-fluorapatite, and within the pores of the bone, sparry calcite occurred, while the channels through which the blood vessels coursed and superficial cracks were filled with a siliceous siltstone cemented by microcrystalline calcite. In another study, David Martill (1987) reported at least seven distinct mineral phases within the bones of Middle Jurassic marine reptiles of the Oxford Clays of England. In addition, in 1993, British researchers Jane Clarke of Hampshire and Michael Barker of the University of Portsmouth reported on the diagenesis in *Iguanodon* bones from southern England in the Wealden Group, Isle of Wight. They demonstrated explicitly that the precipitation of minerals in the pore spaces of bones is dependent on the composition and rate of flow of the pore water of the sediment, which is affected by the permeability and composition of the surrounding sediment.

In an as yet unpublished study of the chemical alteration of bones of *Dryosaurus lettowvorbecki* from the Tendaguru beds of Tanzania, Ethiopian-born Keddy Yemane of the University of Pennsylvania and I found that, in most of the bones examined, the marrow cavity was occupied by clumps of fine- to medium-grained silts of quartz, feldspars, and clays. Careful inspection of some samples showed a lunar-shaped precipitation of prismatic calcite, which suggests that the bones may have rolled around in the mud before consolidation. We also observed multigenerational carbonate precipitation, which suggests successive periods of ground water circulation. The clay and micrite calcite in the marrow cavity indicates clastic infiltration during burial. We also found that the infilling within the pores and voids of the bone were mainly calcite and pyrite, and an admixture of clays, with occasional feldspars. Scanning electron microscopy backscatter elemental maps of a transverse section of a femur of *Dryosaurus* shows the preferential enrichment of certain elements in areas of the bone (Fig. 1.8). For example, calcium is distributed throughout the bone but appears to be concentrated within the channels in the bone, while sulfur is absent in these spaces and appears in larger quantities in the interchannel regions of the bone (Fig. 1.8).

Infiltrations in Fossil Bones

Garland (1987) describes infiltrations "as extraneous material within the bone substance itself." Although apatite remains relatively unaltered, several minor elements can be incorporated into the crystal structure or can form part of the mineral phase filling the voids and fractures in the bone (e.g., Parker and Toots 1970, 1980). Elevated concentrations of barium, sodium, lead, strontium, and zinc have been identified in fossil bones and suggest an affinity for the apatite. In addition, lanthanides also appear to have a strong affinity for phosphates and are located in the bone wall of fossils. In a study of *Gallimimus bullatus* from the Cretaceous beds of the Gobi Desert, Polish researchers Pawlicki and Bolechal (1987) found lanthanium, yttrium, neodymium, and, on occasion, europium. Uranium is commonly found in dinosaur bones from some formations

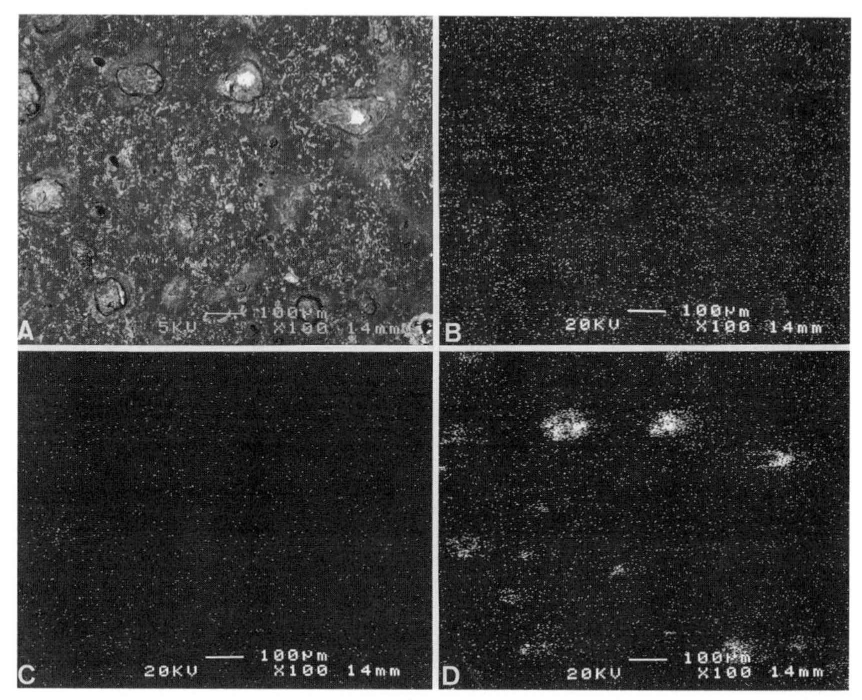

Fig. 1.8. Elemental distribution in the bone wall of *Dryosaurus lettowvorbecki*. *A*. Backscatter image of a transverse section of a femur and elemental maps of *B*, sulphur; *C*, iron; *D*, calcium.

such as the Upper Morrison Formation of the western United States. Zofia Kielan-Jaworowska and J. Pensko (1967) also reported on radioactive bones from the Upper Cretaceous of Mongolia.

Recent work by Trueman and Tuross (2003) has showed that many trace metals (particularly rare-earth elements) that are adsorbed onto bone crystallite surface of the fossil bone reflect the early depositional environment. Therefore, bones that fossilize together should maintain the same trace element signal. This finding could potentially play an important role in establishing provenance of fossils.

Iron in Fossil Bones

Finding iron in a fossil bone is intriguing because iron occurs naturally within the blood and in the environment. Ichthyosaur paleontologist Chris McGowan (1991) speculated that the high concentrations of iron within the spaces once occupied by blood vessels in a hadrosaur bone were derived from the blood of the animal. However, it is highly unlikely that the iron in fossil bone derives from hemoglobin in the blood because the total body iron concentration of various animal species varies between 25 and 75 mg/kg (Finch et al. 1979), while the iron content of hemoglobin is a mere 0.355%, of which only 3–4 mg occurs in the blood plasma. In our study of *Dryosaurus*, Keddy Yemane and I found that iron showed a localized distribution in the intervascular spaces and was not restricted to the spaces of the vascular channels (Fig. 1.8). This finding suggested that it is more likely that the iron is absorbed from the environment. It also concurs with the deduction reached by Polish researcher Roman Pawlicki (1983) that the high concentration of iron in 80 million-year-old *Tarbosaurus* bone probably derived from outside the body of the animal.

In 1997, David Martill and David Unwin (now in Berlin at the Humboldt Museum) described small, mineralized spherules in a Lower Cretaceous archosaurian limb bone (which was possibly a tibia of a large pterodactyloid). They looked very much like the spheres that Mary Schweitzer of the Museum of the Rockies in Montana and her colleague Raul Cano of California Polytechnic State University (1994) described as possibly red blood cells in vascular canals of a tyrannosaur bone. Martill and Unwin (1997) inspected the sample under scanning electron microscopy and using EDAX analysis found that they were positively identifiable as pyrite framboids. British researchers Clarke and Barker (1993) reported on spherical structures from dinosaur bones from the Isle of Wight, which were found to be composed of siderite and barite. These findings raised doubt about whether the similar spheres reported by Schweitzer and Cano (1994) were genuine nucleated blood cells or were diagenetic in origin.

Organic Matter in Dinosaur Bones

Although most organic components decompose during the fossilization process, there have been several reports of proteins and fatty acids preserved within the skeletal remains (i.e., bones and teeth) of dinosaurs. For example, in 1977, Roman Pawlicki identified glycoproteins in an 80 million-year-old dinosaur bone of *Tarbosaurus battaar* from the Gobi Desert using a highly sensitive method of detecting Periodic Acid Schiff reaction (PAS) positive substances, using fluorescent Schiff's reagent. More recently, using immunological assays, Gerard Muyzer of Leiden University and his co-workers (1992) identified the remains of osteocalcin (OC), a noncollagenous bone matrix protein that is strongly bound to the bone mineral by γ-carboxyglutamic (Gla), in the bones of dinosaurs and other fossil vertebrates. They found that antibodies developed against OC from modern vertebrates showed strong immunological cross-reactivity with modern and relatively young fossil samples and significant reaction with some Cretaceous dinosaur bone extracts. The presence of OC was further confirmed by the detection of Gla acid, which is a specific chemical marker for OC in fossil bones because it is absent from invertebrates and microbes, which are the most common sources of contamination. Gurley and colleagues from the Los Alamos National Laboratory in New Mexico, including dinosaur paleontologist David Gillette (1991), successfully extracted proteins from 150 million-year-old *Seismosaurus* vertebra. Although the actual identity of the proteins has not yet been determined, the researchers suggest that they may be from noncollagenous protein from the dinosaur bone or possibly from contaminants in the environment.

One of the most enigmatic proteins that paleontologists would like to obtain from fossilized bone is DNA. Because amino acids have been occasionally sequenced, it seemed possible that those from DNA could also be preserved. One of the intrinsic biases of phylogenetic assessments in paleontology is that they are based on morphological characteristics. In modern taxa, morphology alone is, in general, not significant among many closely related groups. Molecular phylogenies (i.e., those based on nuclear and mitochondrial DNA) are far better. Indeed, the recent work on the evolution of whales has clearly highlighted the problem regarding the use of morphological/paleontological data from the fossil record and that of DNA studies: studies of fossil whales suggest that the ancestor of whales are extinct ungulates called *mesonychids,* but molecular data indicate that hippos (that arose from artiodactyls, not mesonychids) are the closest relatives of whales.

Thus, obtaining DNA of extinct animals could have enormous potential in reconstructing their relationship to extant taxa. In the last decades, molecular techniques for sequencing genes from ancient animals advanced to such an extent that the possibility of recovering ancient DNA appeared to be within reach. Since the first announcement of DNA from a 15–20 million-year-old fossil plant was reported in 1990, a series of other reports have been made.

More recently Krings and colleagues (1997) reported the recovery of mitochondrial DNA from a humerus of a Neanderthal fossil dated at about 40,000–50,000 years ago. Their findings showed clearly that Neanderthals are much more distantly related to modern humans than previously supposed. The amount of difference between modern humans and the Neanderthal suggests that the divergence between *Homo* and *Neanderthalensis* occurred at between 690,000 and 550,000 years ago. Following this breakthrough in obtaining prehuman DNA, DNA was recovered from a 29,000-year-old rib of a baby Neanderthal skeleton found in Russia's Caucasus Mountains (Ovchinnikov et al. 2000). The research team from Scotland, Sweden, Russia, and the

United States found that this DNA showed a mere 3.5% difference from the DNA recovered from the 1997 study and supported the findings that Neanderthals are our cousins rather than ancestors.

DNA has been recovered from a number of animals from the recent past (e.g., from the hides of the animals, Egyptian mummies, and from dry well–preserved archaeological bones and tissues, including frozen corpses (e.g., from 50,000- to 100,000-year-old mammoth remains preserved in the Siberian peninsula). Experiments in laboratories suggested that DNA would not usually survive more than 10,000–100,000 years. However, scientists speculated that perhaps under exceptional circumstances DNA would remain intact for longer periods. DNA needs to be free of water to be preserved, and oxidation should be inhibited. Preservation in amber seemed to offer an optimum environment for harboring ancient DNA because amber rapidly entombs the organism and dehydrates the organic matter.

One of the major problems with reports regarding ancient DNA is that only scraps of DNA have been recovered, which consists of a few hundred of base pairs and have had irreproducible results. This is true of the DNA from Golenberg and colleagues, Miocene *magnolia*, as well as for that of the Cretaceous dinosaur bone (Woodward et al. 1994) and egg findings (Li et al. 1995). In most cases, contamination appears to be a major problem in the DNA amplification technique of polymerase chain reaction. The putative dinosaur DNA was found to match human pseudogene sequence, while that from the dinosaur egg was found to represent sequences of either plant or invertebrate. In our write up about the wood fragment found in the marrow cavity of a dinosaur bone, my colleagues and I (Chinsamy et al. 1996) pointed out that the wood fragment was not visible macroscopically and was only identified as such by the fortuitous thin sectioning. We raised the concern that such an inclusion could have had dire consequences for ancient DNA studies; that is, if this sample had been pulverized for DNA extraction, the extreme sensitivity of the polymerase chain reaction technique could have resulted in the amplification of DNA templates of either or both the wood and the dinosaur. (Normally the specificity of the PCR should preclude amplification of the wood DNA; however, when amplifying ancient or degraded DNA, the annealing temperature is lowered to a level where this specificity is compromised.)

Reports of DNA recovered from insects preserved in amber are now also treated with skepticism because the findings have never been replicated in independent laboratories. Doubt has also been cast by Austin and colleagues (1997), who attempted to recover DNA from several insects preserved in 25–35 mya Dominican amber and from insects in a few-hundred- or thousand-year-old amber and found that he could not recover indisputable insect DNA from such young material. Thus, all DNA recovered from insects preserved in amber is contested, and the oldest authentic DNA is that recovered from frozen carcasses of woolly mammoths.

It is quite extraordinary that proteins (such as osteocalcin) have been recovered from dinosaurs, but given the authenticity problems associated with DNA recovered from relatively recent fossils, I doubt that authentic DNA from dinosaurs will be discovered. Of course it is more engaging to entertain the possibility that such a discovery is yet to come!

Soft Tissue Preservation in Dinosaurs

Under exceptional circumstances, soft tissue can be preserved by, for example, natural mummification, which involves desiccation or freezing (which inhibits putrefaction, i.e., the bacterial breakdown of protein under anaerobic conditions) or by burial in hot-dry environments. Impressions of dinosaur skin have been known since the turn of the last century. However, in recent years, some incredible revelations of dinosaur soft tissue have been reported. The spectacular burial ground of the Lower Cretaceous lacustrine limestones of Las Hoyas in Spain has yielded a diverse biota of plants, insects, crustaceans, fishes, and birds, including feather impressions of birds. Phosphatized muscle tissues of fishes and decapods from this locality have been reported. Recently, Derek Briggs of Bristol University, Philip Wilby of the British Geological Survey in Nottingham, and Spanish scientists Bernardino Pérez-Moreno, José Luis Sanz, and Marian Fregenal-Martínez described mineralized soft tissue of the ornithomimosaur, *Pelecanimimus polydon* (Briggs et al. 1997). The researchers surmise that the carcass fell into a freshwater lake at Las Hoyas and sank into the mud. A bacterial mat then grew over the carcass and abetted the deposition of fossilizing minerals by entrapping phosphorous released from the decaying carcass. This phosphatized microbial mat that covered the carcass preserved the outline of this dinosaur. "Integumentary impressions" in the region of the throat and near the ribs and elbow have been identified as mineralized traces of soft tissue that preserved replicas of the original skin and muscle tissue as iron carbonate. Impressions on microbial mats that enshrouded the carcass show that *Pelecanimimus* had a pelican-like throat pouch and a soft occipital crest. A detailed mold of the skin was also preserved and revealed a wrinkled surface without any enlarged scales or feathers.

Brazilian paleontologist Alexander Kellner, a well-known expert on pterosaurs, reported in *Nature* (1996) on the mineralized, preserved soft tissue in a small theropod from the Lower Cretaceous Santana Formation of Brazil. Kellner removed a sample of

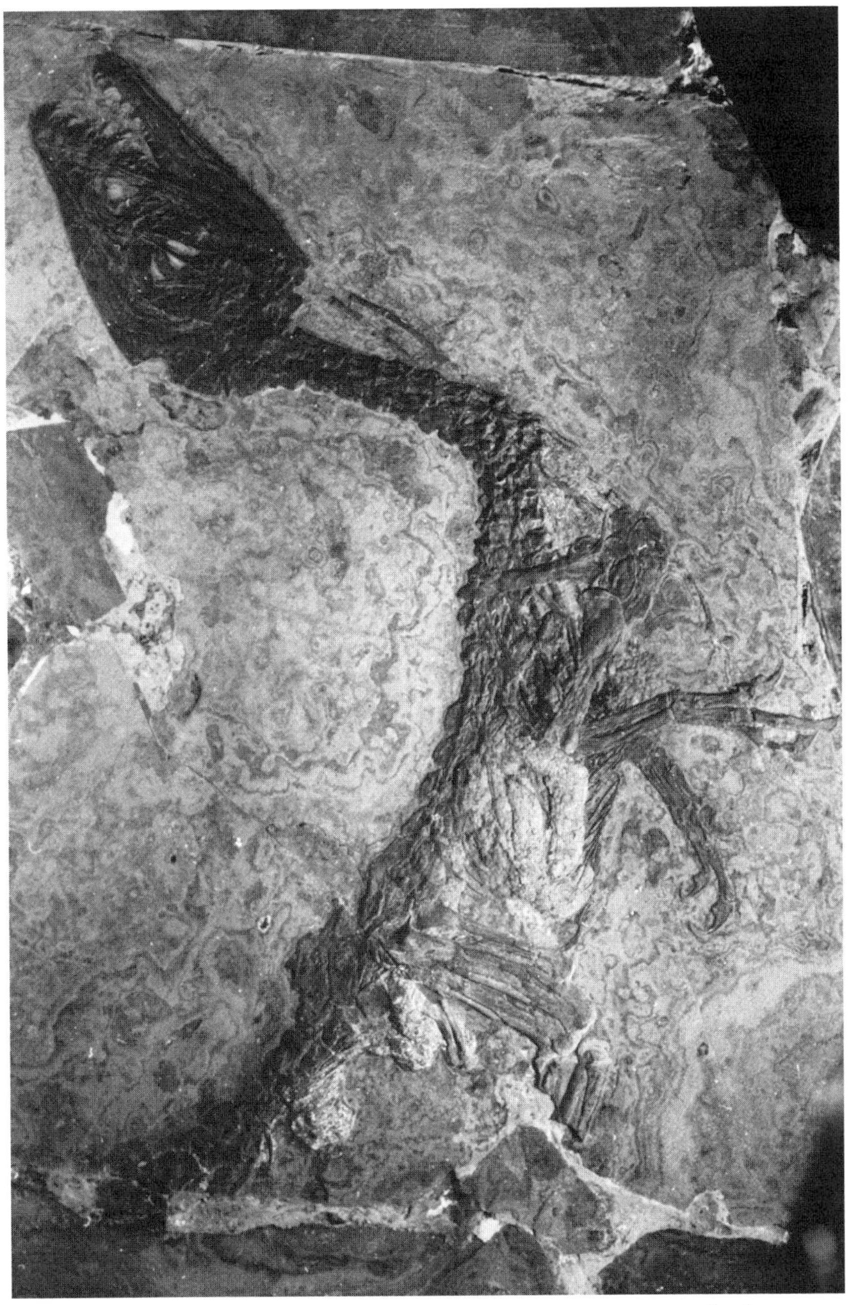

Fig. 1.9. *Scipionyx samniticus,* the baby dinosaur from Italy that preserves parts of its large intestine, trachea, and some muscle tissue. Skull length equals 55 mm. Image courtesy of Willem Hillenius.

the soft tissue preserved in the region of the femur-tibia and fibula and examined it under the scanning electron microscope. The analysis revealed that the soft tissue represented striated muscle fibers and was preserved as calcium phosphate. Quite remarkably, some of the fibers preserved actual striations of the muscle tissue. Kellner also speculated that the mineralizations filling the channels of transverse and longitudinal sections of a fragment of the femur possibly represented blood vessels that penetrated the bone.

When the first Italian dinosaur was described by Cristiano Dal Sasso and Marco Signore from the Museo Civico de Historia Naturale in Milan and Universita degli Studi di Napoli in Naples, respectively, it caused quite a stir, not simply because it was a 113-million-year-old baby dinosaur from Italy but because of the unique preservation of the specimen. *Scipionyx samniticus,* as it was named, perfectly preserves the trachea, major parts of its intestines, colon, and parts of its liver (Fig. 1.9). The specimen was preserved in a fine-grained limestone that formed in a shallow, oxygen-deficient lagoon, which is probably why no scavengers disturbed the carcass. Exceptionally well-preserved fossilized fishes and invertebrates with soft tissue are also known from these deposits.

An intriguing recent soft tissue discovery is that of the "heart" of the 66-million-year-old *Thescelosaurus,* a 29 kg, 4-m-long herbivore that was described by Paul Fisher of North Carolina State University and his colleagues (Fisher et al. 2000). Although the specimen was only described in April 2000, Michael Hammer and his son Jeff discovered the specimen in 1993 in the Hell Creek Formation in South Dakota. Now, several years later, x-ray computed tomography (CT) scans of the dinosaur's chest region and three-dimensional composites revealed what appear to be a four-chambered heart with an arched single aorta. Taphonomic indicators suggest that the specimen was buried in wet, oxygen-free sand and that instead of decaying the soft tissue were saponified (i.e., turned into a soaplike substance and petrified). The specimen also shows well-preserved tendons that connect to the spine and cartilage on its ribs. However, the "heart" has not won over many paleontologists, who are, in general, skeptical about it, and believe that it is simply a concretion.

Conclusions

Bone of all vertebrates comprise the same basic components that are organized in different ways, depending on a host of factors, such as phylogeny, ontogeny, the habitat in which it lives, and the local environment. During the fossilization process, organic components of the animal rapidly decompose, and soft tissue rarely is preserved.

However, the mineralized components of the skeleton (i.e., bones and teeth) resist decay. The intimate association of the organic (collagen) phase and the inorganic (mineral) phase, as well as the minute size of the crystallites of the apatite and their location within the collagen structure, endows bone with durability in the fossil record. This permits researchers to compare the bone microstructure of dinosaurs with that of other vertebrates to derive information about various aspects of dinosaur biology.

CHAPTER TWO

Dinosaur Bones Discovered

Finding a dinosaur bone or, for that matter, even just a scrap of a dinosaur bone or tooth is thrilling. Depending on the bone and its preservation, more often than not one can work out exactly which part of the skeleton is represented, and if diagnostic characteristics are preserved, it is possible to determine to which animal it belonged. The questions then become, What did this animal look like? Was it big or small? What sort of biology did it have? How did it die? Regrettably, such aspects of an animal's life that can be observed readily in a living animal are not preserved in the fossil record.

Once the skeletal remains are excavated and prepared out of the surrounding matrix, it is essential to assess how the specimen is related to other dinosaurs and to reconstruct and understand it as a living creature. In the past, anatomical similarity seemed to suggest phylogenetic relationships (i.e., that the changes between ancestor and descendant were thought to be recognizable in a particular time span). This model assumed that the fossil record is complete, but in reality, it is not. Because of this and largely subjective interpretations, this type of phylogenetic assignment is far from adequate. Cladistic phylogeny is now accepted as a more reliable, objective method that makes fewer assumptions about the completeness of the fossil record. It does not simply use all similar characteristics but also considers the evolution of novelties. Cladistics has been rigorously applied to unravel the phylogenetic relationships of dinosaurs.

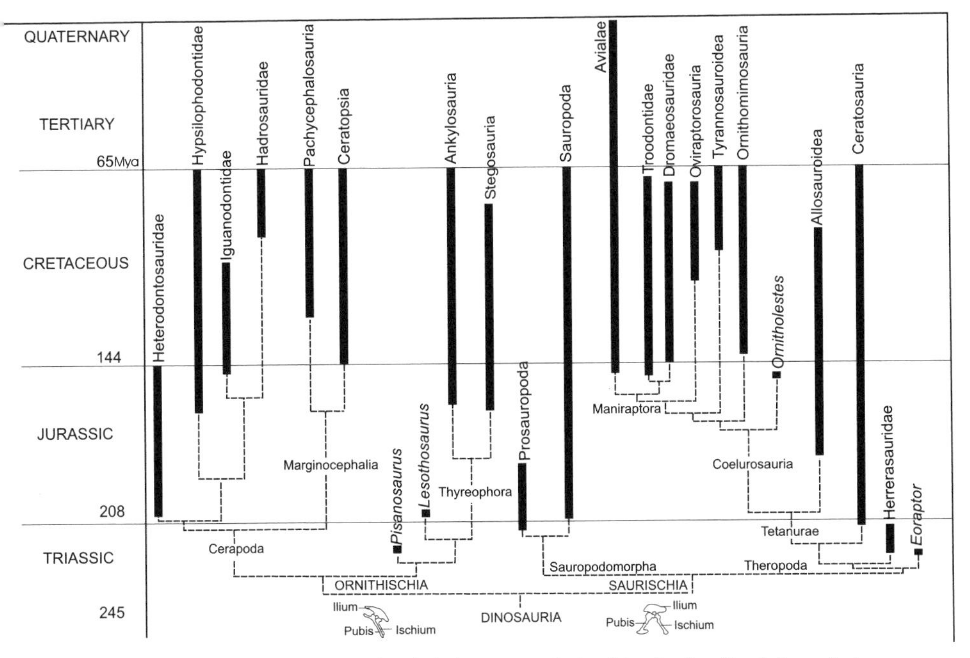

Fig. 2.1. Dinosaur diversity and postulated phylogenetic relationships (broken lines) through time. "Mya" refers to millions of years ago (modified from Sereno 1999; Benton 2000).

Synopsis of Dinosaur Phylogeny

In 1887, British paleontologist Harry Govier Seeley distinguished two major groups among the Dinosauria based on the structure of the pelvic girdle. He recognized the saurischian, or reptile-hipped, dinosaurs as having a three-pronged pelvic girdle in which the pubis is directed forward and the ischium is directed backward, and the ornithischian, or bird-hipped, pelvic girdle in which both the pubis and ischium were directed backward as in modern birds (Fig. 2.1). Though the field of dinosaur studies has grown substantially, Seeley's 1887 distinction of saurischian and ornithischian dinosaurs based on the structure of the pelvic girdle has stood the test of time and is still valid. Saurischian dinosaurs comprise both herbivores and carnivores, while the ornithischian dinosaurs are all herbivores.

The Ornithischian Dinosaurs

During the Late Triassic, ornithischians were relatively rare, but by the Early Jurassic, the major ornithischian clades were established. The two main groups of the ornithischian dinosaurs are the Cerapoda and Thyreophora.

The most diverse of the ornithischians, the Ornithopoda, consist of the heterodontosaurids, hypsilophodontids, iguanodontids, and hadrosaurids. The heterodontosaurids first appeared in the Early Jurassic, and even though they have prominent canines, and grasping forelimbs equipped with claws, they show adaptations for herbivory. The most conservative of the Ornithopoda, the hypsilophodontids, enjoyed a long history from about the Middle Jurassic to the end of the Cretaceous. In contrast, the iguanodontians underwent substantial changes in their evolution from the Late Jurassic to the Early Cretaceous, and in the Late Cretaceous, the duck-billed hadrosaurs radiated into a diverse array of forms.

The marginocephalians have a distinctive bony shelf on the posterior margin of the skull and occur exclusively on the northern continents and mainly from the Upper Cretaceous of western North America and Asia. Two main groups, the pachycephalosaurs and the ceratopsians, are recognized. The pachycephalosaurs have distinctively thickened skull domes. The ceratopsians, however, are the frilled and horned dinosaurs that characteristically show variation in the number and in the position of the horns, as well as in the size and orientation of a bony frill comprising the parietal and squamosal bones. They were a large group of dinosaurs with representatives found mainly in the mid-Cretaceous of Mongolia and the Late Cretaceous of North America. The horns and frills probably functioned in defense and/or to assert dominance.

The ankylosaurs and stegosaurs constitute the Thyreophora. The earliest and most primitive stegosaurs, such as the Middle Jurassic Chinese *Huayangosaurus,* have reduced lateral osteoderm rows, while the pair along the dorsal midline has larger plates over the neck that grade into pointy spines along the tail. More derived stegosaurs have expanded the dorsal bony plates that probably functioned for defense, for display, and/or for thermoregulation. When compared to their earlier relatives, stegosaurs also have longer hindlimbs versus their forelimbs and low narrow skulls. The Ankylosauria are heavily armored forms with massive bony protuberances at the ends of their tails.

The Saurischian Dinosaurs

The sauropodomorphs constitute a significant radiation of herbivorous dinosaurs. By the Late Triassic, the sauropodomorphs separated into two distinct groups, the prosauropods and the sauropods, though there is some debate about whether the prosauropods were ancestors to the sauropods. The prosauropods rapidly diversified to become the dominant large-bodied herbivores from the Late Triassic through to the Early Jurassic and showed remarkable conformity in their design despite their wide distribution across Pangaea. Sauropods, the long-necked giants of the Mesozoic, were relatively rare in the Early Jurassic when the ornithischians diversified but underwent a major radiation after the prosauropods became extinct in the Middle Jurassic. Among the sauropomorpha, giant size seems to have evolved very quickly. The early prosauropods, such as *Plateosaurus,* were only about 7 m in length, but by the Early Jurassic, the sauropodomorphs, such as *Melanorosaurus,* had reached lengths of about 12 m and body weights of about 10 tons; size continued to increase until the Late Jurassic when the largest known sauropods evolved. Michael Benton (1989), a well-known British paleontologist, has calculated that this gigantic leap from 10 tons to about 80 tons occurred in about 20 million years. There is no clear reason why the sauropods evolved into giants. However, Indiana paleobiologist Jim Farlow has speculated that perhaps a range of factors such as the greenhouse conditions of the Mesozoic and low food consumption rates permitted the evolution of dinosaur gigantism.

A distinctive characteristic of all theropods, including birds, is that they are obligatory bipeds. The earliest theropods, *Eoraptor* and *Herrerasaurus,* show distinctive predatory adaptations such as a flexible jaw joint and an elongated hand with three functional digits adapted for grasping and raking. The ceratosaurs were the major group of theropods during the Late Triassic and Early Jurassic. By the Middle Jurassic, the tetanurans had diversified on all continents and split into the allosauroids and the coelurosaurs, from which birds are considered to have descended. The oldest tetanuran is the Early Jurassic crested allosaurid, *Cryolophosaurus* from Antarctica, which is very similar to the Upper Jurassic allosaurids from several other continents. During the Cretaceous, the allosaurids grew to massive sizes, comparable to the tyrannosaurids. The nonavian coelurosaurs include a diverse array of small to medium-sized predators and include the tyrannosaurs, troodontids, dromaeosaurs, and oviraptids.

Paleoecological Analysis

A paleoenvironmental analysis of the depositional environment of the fossils provides information regarding the habitat preference and behavior of the once-living animal. For example, dinosaur remains are known from every continent in the world, including high-latitude environments of Alaska and Antarctica, which implies that they were capable of living under a variety of conditions at different latitudes. Paleoenvironmental analysis of localities where dinosaur remains have been found shows that they frequently preferred moist, lowland environments, although they are also known to have lived in shallow marine areas through to arid, Aeolian deposits (Farlow et al. 1995). Sauropod genera of the Late Jurassic Morrison Formation of the western United States are found across a wide range of paleoenvironments over a distance of 1000 km. However, the Morrison Formation stegosaurs suggest that they preferred a drier habitat. Other studies also show that different dinosaur genera may have had different habitat preferences.

Most dinosaur remains consist of articulated, partial skeletons and, in many cases, constitute fragmentary remains. Several discoveries have identified more than one dinosaur from a single dig site (locality). Sometimes many individuals of a particular species (i.e., a monospecific assemblage) are found, which suggest that these dinosaurs lived in herds or packs of single species, and they were together as a group when some catastrophe befell them. Several dinosaurs (e.g., ceratopsians and hadrosaurs of Alberta) are known from monospecific bone beds that suggest herding behavior. It is also apparent that some predatory dinosaurs, even including the largest of them all, lived and hunted in social groups. Aside from isolated finds and monospecific dinosaur assemblages, discoveries of multiple dinosaur taxa in a single locality strongly indicate pack-hunting behavior of predatory dinosaurs. Most famous among these is John Ostrom's discovery of four meat-eating *Deinonychus* with the herbivore *Tenontosaurus* at the Shrine site locality in Montana's Cloverly Formation.

On occasion, telltale indications of cause of death are visible on the skeletal remains in the form of gorge, puncture, or gnaw marks caused by the teeth of predators or scavengers (e.g., tooth puncture marks, measuring about 1 cm, have been recognized in a triceratopsian pelvis, which Greg Erickson and colleagues [1996] from Stanford have determined were caused by *Tyrannosaurus rex*). Sometimes the actual teeth of a predator are found embedded or associated with the skeleton. The ceratopsian graveyards of Dinosaur Provincial Park in Alberta represent another death assemblage that clearly shows that the ceratopsians lived in herds of about 300 individuals and that they were stampeded to death as they attempted to cross a flooded river.

Reconstructing Dinosaur Biology

Once the taxonomic affinities of the dinosaur are established, the challenge for paleobiologists is to assess various aspects of the animal's evolutionary biology, which often refers to unpreserved soft tissues. In the past, there was much controversy about how soft tissues of extinct animals were reconstructed. However, today paleobiologists rely on the more thorough extant phylogenetic bracket (EPB) approach proposed by Ohio University anatomist and dinosaur paleontologist Larry Witmer (1995a, 1995b). The methodology provides paleobiologists with a rigorous phylogenetic means of reconstructing the soft anatomy (as well as other behavioral or functional traits) of extinct animals from the observed soft tissue anatomy of its closest living relatives. The selection of appropriate, most closely related, and more distantly related extant taxa that forms the extant phylogenetic bracket is important because the character assessment in these extant taxa are crucial in determining the ancestral condition in the fossil clade (Fig. 2.2). The second more distantly related extant taxon serves as the outgroup to the first two, allowing an estimation of the ancestral feature at the outgroup node. By reorganizing the extant outgroups to the periphery (by rotation of the cladogram node) so that the two extant taxa flank the fossil taxon, a graphical depiction of the extant phylogenetic bracket of the extinct taxon is obtained (Fig. 2.2). The extant taxa form a "bracket" because they constrain any inferences about the fossil taxon.

Thus, the character assessment of the most recent common ancestor of the fossil taxon and its first extant outgroup (i.e., its sister taxon) is critical for determining the ancestral condition of the fossil clade. When there is character variation between the two chosen outgroups, often examination of the particular trait in additional living outgroups helps resolve the problem. The EPB is therefore based on biological homology, and it allows a way to establish limits of inferences and permits a hierarchy of inferences to be established. Witmer (1995a) also stresses that for a particular soft tissue to be reliably attributed to a fossil taxon homologous soft tissues must be shown to produce homologous osteological correlates (i.e., unambiguous bony features on the bones caused by known soft tissue components). It is therefore evident that the EPB approach enables researchers to distinguish between sound and unconfounded soft tissue reconstructions in fossil taxa.

In a straightforward case, all similarities between the two extant taxa can be hypothesized to have been present in the common ancestor and in all its descendants, including the fossil taxon. If osteological correlates of the soft tissue are present in the fossil taxa, it is quite reasonable to infer the presence of the soft tissue, and the hypothesis survives the so-called congruence test (whereby the homology characterizes the monophyletic group; Witmer 1995a). Because the hypothesis suggests that all other descendants of the common ancestor should have the osteological trait, examination of fossil taxa would test the hypothesis. If all the fossil taxa have the particular trait, it is reasonable to infer and to reconstruct the soft tissue in the fossil taxa. However, when one or more clades of fossil taxa lack the osteological characteristics, parsimony would help to formulate a sound hypothesis. The situation is complicated when one or both of the extant outgroups lack the specified soft tissue. In this case, higher levels of inferences are needed, and the deductions made are more speculative and should be treated with caution.

The EPB methodology, therefore, provides a rigorous way to assess soft tissue and other behavioral attributes of fossil taxa and has been applied extensively to understand dinosaurs as once-living animals. Considering this approach, it is obvious that to reconstruct reliably various aspects of soft tissue anatomy (as well as other behavioral or functional traits) of dinosaurs, it is crucial that the soft tissue anatomy of their closest living relatives (i.e., birds and crocodiles) are assessed (Fig. 2.2). If the soft tissue and its osteological correlates are present both in birds and in crocodilians, it is reasonable to assume that the soft tissue was also present in their common ancestor and all their descendants (i.e., including dinosaurs). However, if the trait is

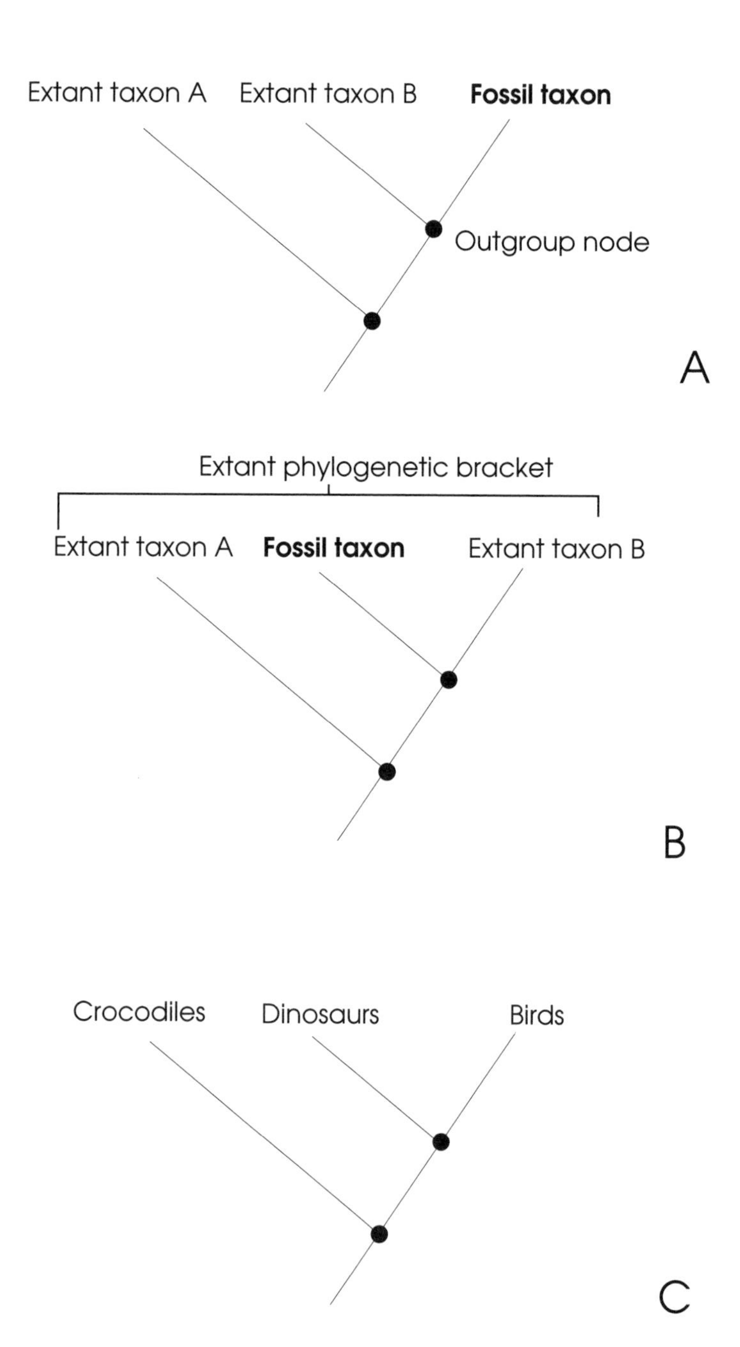

Fig. 2.2. *A*. Phylogenetic relationships of a fossil taxon and its first two extant sister taxa. *B*. Rotation around the outgroup node in *A* brings the extant taxa to the periphery to depict graphically the extant phylogenetic bracket. *C*. Extant phylogenetic bracket of dinosaurs. *A* and *B* are redrawn from Witmer (1995b).

present in only one extant taxon, then caution should be exercised when inferring the characteristic to dinosaurs.

Macroscopic Studies of Dinosaur Skeletons

By studying the gross anatomy of the fossil bones, one can reasonably reconstruct the overall shape and size of the extinct animal. Determining whether the animal was bipedal or quadrupedal is easily deduced. The morphology of the bones and teeth, skeletal proportions, and joint articulations can be, in general, readily assessed from dinosaur skeletons, and together with studies of muscle scars on the bones (and comparisons with the extant phylogenetic bracket), deductions can be made regarding various biomechanical and functional aspects of the skeleton. Comparing the dinosaur to its closest morphological, ecological, and behavioral equivalents is also very useful. For example, tails, clubs, spikes, horns, armor, and so on, among modern vertebrates are used for intra- and interspecific behavioral interactions, such as threat, combat, territoriality, submission, and dominance. It is reasonable to assume that the presence of such skeletal features in a dinosaur skeleton functioned similarly. Many functional interpretations of skeletal material are often supported by other lines of evidence—for example, the erect posture of dinosaurs is confirmed by the narrow "gauge" of dinosaur tracks (Farlow et al. 1995); gut contents or coprolites (fossilized dung) can often corroborate herbivorous or carnivorous dinosaurs identified on the basis of tooth structure (Farlow et al. 1995).

Because dinosaurs are amniotes, even in the absence of fossil evidence, it would be reasonable to assume that they possessed a heart, blood vessels, lungs, and stomach. Deductions about the type of heart structure, the type of blood circulation, or type of lungs they had can be made by considering their extant phylogenetic relatives (i.e., crocodilians [Crocodylia] and birds [Neornithes]). It is reasonable to infer the presence of soft tissues in dinosaurs because they are present both in crocodilians and in birds. However, as previously explained, when a structure is present only in one of the crown groups (e.g., a diaphragm in crocodilians and the absence thereof in birds or a sacral glycogen body in birds but not in crocodilians), inferences of the presence of the structure in dinosaurs becomes complicated (Farlow et al. 1995; Witmer 1995a). It is even more speculative to deduce structures in dinosaurs (e.g., carotid hearts) when neither of the crown groups have such structures (see Farlow et al. 1995).

Recently, Larry Witmer (2001), applying the EPB approach to assess soft tissue in dinosaurs, studied forty-five animal species of birds, dinosaurs, and lizards to infer the location of flesh around the bony nares of dinosaurs. Witmer painted the fleshy nostrils with a liquid visible under x-rays, and then compared this position with that of the actual bony openings in the skull. He found that in nearly every case the fleshy nostril was located frontally near the mouth. Thus, it appears that for most of their

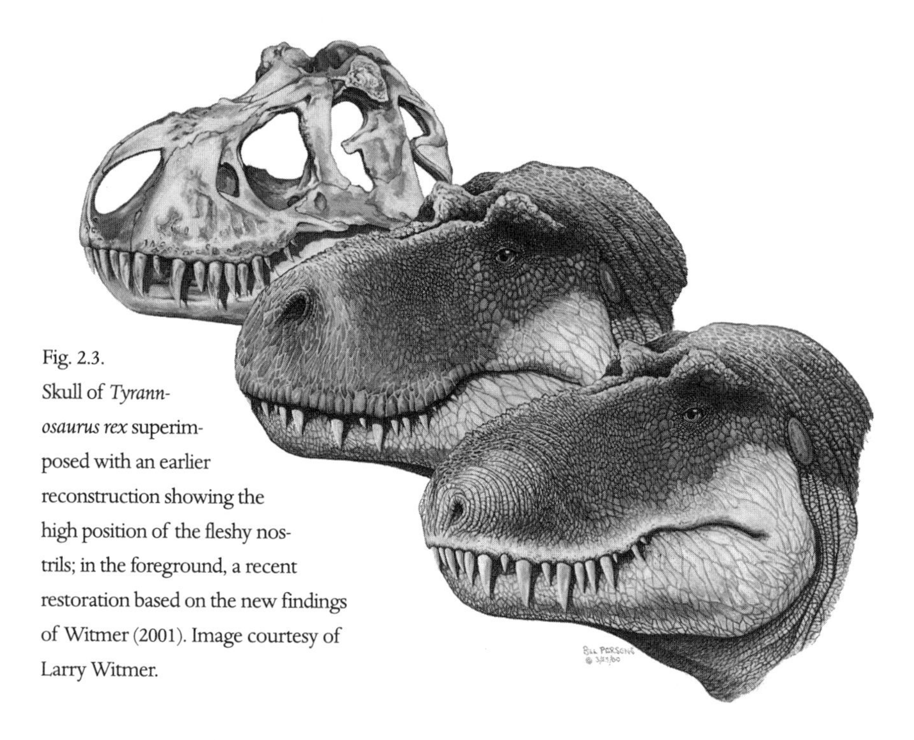

Fig. 2.3.
Skull of *Tyrannosaurus rex* superimposed with an earlier reconstruction showing the high position of the fleshy nostrils; in the foreground, a recent restoration based on the new findings of Witmer (2001). Image courtesy of Larry Witmer.

history dinosaurs have been incorrectly reconstructed, with nostrils fairly high at the back of the bony nostrils (Fig. 2.3). It is possible that the earlier reconstructions were based on the outdated idea that dinosaurs were amphibious and therefore would have required nostrils that were high and toward the back of the snout (so that the animal could have breathed while almost submerged in water). The analysis by Witmer (2001) is now aligned with the current ideas of dinosaurs as terrestrial. The location of the fleshy nostrils at the front of the skull is also supported by a number of grooves in the skull, which could be evidence of a network of blood vessels generally associated with the fleshy nostrils. A consequence of more forwardly placed fleshy nostrils is that the nasal passage is lengthened, which would be important in conserving warmth and fluid from exhaled or inhaled air. It is also possible that the lengthening of the nasal passage influenced the animal's sense of smell.

Besides using the EPB for deducing soft tissues of dinosaurs, it has also helped immensely in constraining speculations about various aspects of dinosaur behavior. For example, aspects of nesting behavior, such as incubation of the eggs and whether dinosaurs cared for the newly hatched young, cannot be readily deduced from the fossil record. However, by considering the extant phylogenetic bracket, the range of nesting behavior of crocodilians and birds can be ascertained and more reliable behavioral characteristics can be ascribed to dinosaurs. As with osteological traits, if the behavioral characteristic is present in both crown groups, it can be reliably inferred about dinosaurs. However, when it is present in either crocodilians or birds, deductions require higher levels of inferences and must be treated with caution.

CT Scans of Dinosaur Bone

X-ray computed tomography (CT scanning), which produces two-dimensional "slices" through objects, have substantially contributed to the study of dinosaur skeletal material. The technology is particularly useful because it allows a researcher to study internal areas of fossil bones without having to destroy the fossil by serial sectioning (a destructive procedure that was frequently used in the early days to study the internal structure of braincases). Thus, CT scanning has proved particularly useful in studying the internal morphology of skulls and has enabled studies of the course of nerves and blood vessels inside the braincase. Ruben and colleagues (1996) scanned several dinosaur skulls (*Nanotyrannosaurus, Ornithomimus,* and *Hypacrosaurus*), and after studying the internal morphology, deduced that dinosaurs had relatively narrow cross-sectional areas of the nasal cavity and lacked respiratory turbinates. Advancements in CT scanning technology have led to greater resolution and penetrating power and using other digital morphology software has enabled generation of two-dimensional and three-dimensional visualizations of the skeletal material. Recent analyses of pterosaur skulls (*Rhamphorhynchus* and *Anhanguera*) using CT scans and DigiMorph have led to an incredible amount of information about how they maintained balance during flight, as well as details about the anatomy of their brains (Witmer et al. 2003).

High-resolution CT scans are increasingly used in fossil preparation (i.e., to analyze fossils before preparing the fossil out of the surrounding matrix). This has proved immensely useful in cases where very fragile bones are being prepared, such as those of birds or pterosaurs. Indeed CT scanning technology applied to dinosaur eggs has led to the discovery of several embryos preserved within unbroken eggs.

Tim Rowe, director of the Vertebrate Paleontology Laboratory at the University of Texas, Austin, has been a leading figure in developing the use of high-resolution CT scans in paleontology and has enabled the availability of several two- and three-dimensional visualizations and animations of a variety of fossil vertebrates. While using CT scanning technology a few years ago, Tim Rowe uncovered "Archaeoraptor" as a fraud by showing that it was a composite of a new bird species and a nonavian dinosaur (Rowe et al. 2001).

Recent innovative work by Emily Rayfield and several colleagues (2001) used CT scans to obtain a three-dimensional model of an *Allosaurus fragilis* skull to which finite element analysis was applied to assess its mechanical properties under stress and strain. They found that though the *Allosaurus* skull had a "light," "open" design and a rather weak muscle-driven bite force it exhibited unusually high cranial strength (Rayfield et al. 2001).

It is apparent that macroscopic studies have moved away from traditional anatomical descriptions of dinosaur bones. Today, dinosaurs are best understood within a phylogenetic framework, and by application of the extant phylogenetic approach, soft tissue anatomy can be reasonably deduced. Within this context, new scientific advancements in technologies are increasingly applied to fossilized skeletal material with remarkable effect on our understanding of their biology.

Microscopic Studies of Dinosaur Bones

The external morphology of dinosaur bones is regarded as vital in reconstructing the animal. However, the architecture of the fossil bone and its microscopic structure also contain a host of (often untapped) information about the biomechanical functioning of the skeleton, as well as information about the growth processes that have affected its development. Studies of modern bones show that the material properties of the bone are strongly influenced by the shape of the bone, its cross-sectional geometry, and its bone microstructure.

Studies of the microscopic structure of dinosaur bones rely heavily on comparative studies with modern vertebrate bone, particularly with comparisons of the EPB (i.e., crocodilian and bird bones). Aside from phylogeny, ontogeny, and environmental (seasonality), the internal structure of bones is affected by lifestyle factors, such as large size, cursorial, aquatic, or aerial adaptations. Therefore, bone tissues of morphological, behavioral, and ecological equivalent vertebrates are often included in studies of dinosaur bone tissues.

Many dinosaur paleontologists would cringe at the thought of cutting up precious dinosaur bones. However, because x-rays and CT scans cannot (as yet) discern microscopic structure of bone and because, like modern vertebrate bone, bones of dinosaurs are too dense for transillumination, they have to be thin-sectioned so that the bone microstructure can be analyzed using light microscopy.

One of the most important steps in undertaking studies of dinosaur bone microstructure is selecting the study sample. In the past, isolated bones of dinosaurs were generally sectioned, and though these are still important in documenting the different tissue types among the Dinosauria, these days, bones to be sectioned require a high degree of selectivity. For example, if one is addressing a specific question such as variability in skeletal microstructure, then it is necessary to sample as many bones as possible of the specified dinosaur. However, because the bone-sectioning process for histological analysis is destructive, it is extremely difficult to find museum collection managers that will support such research. To get around this problem, bone bed material with an assortment of anatomical remains are slightly less precious and more easily obtainable and are suitable for documenting the variability of tissue types in the skeleton. However, bone bed material is not ideally suited for comparative studies considering ontogenetic growth because assembling a skeleton from a bone bed is not the equivalent of studying a single individual. In the latter case, all of the bones in the skeleton have been subjected to the same internal and external factors, such as environment, life history, age, and gender.

If the aim of the study is to document how histological characteristics change during growth, then serial sectioning the entire skeletons of differing ontogenetic ages is necessary. Because the possibility of ever obtaining such a sample is unlikely, the next best thing is to constrain the study by looking at long bones of different ontogenetic stages. Long bones, in particular femora and tibiae, are well recognized as providing the best documentation of histological variation through ontogeny. In femora, it is best to examine the midshaft area, which is considered to be the neutral area (i.e., the area least affected by the remodeling processes of the growth of the bone). Making longitudinal sections of the articular regions of the bone is the best way to assess growth in length.

If the aim of the study is to assess histological variation in bone microstructure, then it is imperative that the sample includes as many skeletal elements as possible and that comparisons are also made of different parts of each bone. It may also be a good idea to serial section bones to see how the microstructure varies according to the morphology of the bone.

Thin Sectioning Dinosaur Bone

Because thin sectioning is a destructive process, it is vital that all measurements, photography, and detailed descriptions are made before beginning the sectioning process. If the bone to be sectioned is rather rare, or from a reasonably complete skeleton, it is advisable to make a cast of the bone before sectioning. The cast can be substituted for the original bone in the skeleton so that morphological details are not entirely lost. Once all these preliminaries are done the procedure for thin sectioning can begin.

Even though fossil bones have survived millions of years of burial, they are usually surprisingly fragile and brittle and therefore require special procedures for thin sectioning. The first step in the thin-sectioning process is to embed the bone in a clear resin. The resin penetrates all the natural pores in the bone, as well as the fractures that have resulted during fossilization, and serves to strengthen the bone. There are many different types of resins that could be used. When choosing a resin, it is important to look for one that is clear and that will not interfere with the optical resolution of the bone tissue. It is possible to purchase special embedding vessels (i.e., containers in which to place the bone and the embedding resin), but usually ordinary plastic containers can be used. I suggest that the container is coated lightly with petroleum jelly before pouring in the resin because this will facilitate the removal of the resin block once it has cured (hardened). To ensure good penetration of the bone by the resin, the resin should be allowed to cure in a vacuum environment. This will also remove excess air bubbles. Once embedded in the resin, the fossil bone can be treated in much the same way as a petrographic sample (i.e., rock samples). Geologists will immediately know what I mean by this, but biologists will find the following explanation useful.

The embedded bone is sectioned in the required direction (i.e., transversely, longitudinally, or tangentially) using a saw with a diamond-encrusted blade. The cut surface of the bone is then ground down on increasingly fine, waterproof abrasive discs on a rotating lapwheel or with decreasing particle-size silicon carbide powder (such as Carborundum powder) on a glass sheet. This grinding down removes the scratch marks caused by the blade during cutting. Final polish of the surface is obtained using a special velvet cloth on a lapwheel or manually on a glass sheet using aluminum oxide suspension. Once the cut surface has a high polish, it is mounted onto a glass slide with resin and left to harden. Considering a standard petrographic laboratory, the specimen is then either put through an automatic thin-sectioning machine or cut individually using a vacuum chuck to hold the mounted slide, that is fed to a rotating water-cooled, diamond-encrusted blade. The latter is time consuming, but there is more control over the resulting sections. In the automated process, one has to be careful that each slide is exactly the same thickness and that the amount of resin used to mount the slide is uniform because the micrometer is set to cut all the sections at the same level. In our lab, we section each slide individually using a precision cutter. The result of the sectioning is about a 1-mm-thick piece of bone stuck onto a slide. The cut surface is ground down using the same procedure as above. At various stages during this process, the section is examined under the microscope to ascertain whether the histological details are visible. This is usually acquired at a thickness of about 20–30 microns. Just before obtaining optimal thinness, the section should be polished using a velvet polishing cloth on the lapwheel and "polishing slurry." The thin section of the fossil bone is now ready to be viewed under the microscope. For a more detailed description of the thin-sectioning process, please refer to Chinsamy and Raath (1992).

Since interpretation of fossil bone tissues depends on understanding the factors that affect modern bone tissues, research in fossil bone tissues would often involve thin sectioning bones of modern animals as well.

Thin Sectioning Modern Bones

The microstructure of modern bone can be studied either by examining calcified or decalcified sections of bone. (Decalcification involves the removal of the mineral component of the bone.) Most conventional texts on bone deal exhaustively on various preparation processes for decalcified and stained bone tissues, and I will therefore only briefly mention the tried and tested methods used in my lab.

Bones can be decalcified using various solutions, though I have used a 4%–5% solution of nitric acid (HNO_3) with success. The amount of time required for decalcification is variable and depends on the actual size of the specimen being decalcified, as well as the type of bone and age of the animal. Once decalcification is completed the specimen can either be embedded in a standard paraffin or celloidin medium for sectioning with an ordinary microtome or simply frozen for thin sectioning using a freezing microtome. In my lab, we tend to follow the latter procedure. After cutting, the thin sections can be stained with a number of different stains, though I frequently use Ehrlich's haematoxylin (prepared according to the standard recipe), which is a good stain to use to observe growth marks in bone.

In our lab, we more frequently prepare thin sections of calcified bone tissue. Although most histology books suggest simply grinding down the sawed surface of a cut bone sample, it is usually much better to embed the bone in a polyester resin and then follow the method described above for fossil bone: progressively grinding down the surface (to a thickness of about 30–20 microns) on progressively finer abrasives until the cut surface is polished and the microstructure of the bone is visible under a light microscope.

Once thin sections of fossil and modern bones are obtained, they can be examined using light microscopy (transmitted and/or polarized), which allows a documentation of the tissue structure (i.e., combination of cells and intercellular substances) of bone.

Electron Microscope Studies of Dinosaur Bone

Studies of dinosaur bone using light microscopy were well underway by the mid-1900s. However, electron microscopy analyses lagged behind. By the late 1950s and 1960s, Enlow and Brown (1957, 1958) and Roman Pawlicki et al. (1966) reported on transmission electron studies of dinosaur bones. Pawlicki (1975) also provided the first scanning electron microscopy studies of dinosaur bone. In this landmark study, Pawlicki examined the 80-million-year-old phalanges of *Tarbosaurus battaar* from the Gobi Desert and provided the first spatial structure of a dinosaur bone and details of the orientation and content of the vascular canals in the bone. In 1976, Pawlicki published a detailed description of how to prepare fossil bones for scanning electron microscopy as well as how to isolate osteocyte lacunae from fossil bone. Roman Pawlicki was a prolific publisher, and over the next few years, he wrote about various aspects of dinosaur bone. Apart from providing detailed microstructural and ultrastructural descriptions of the spatial distribution of dinosaur osteocyte lacunae (e.g., Pawlicki 1978), Pawlicki also conducted a host of biomolecular studies on dinosaur bones (see the section "Molecular Studies of Dinosaur Bones").

In 1993, Claudia Barreto and colleagues published a comparative evaluation of the histology and ultrastructure of the growth plates of juvenile dinosaurs from the Upper Cretaceous Two Medicine Formation of Montana with that of extant birds and reptiles. Though the experimental design of the study was flawed in several ways (e.g., comparing different aged extant reptiles with juvenile dinosaurs and birds), the scanning electron microscopy data generated in the study are interesting. Chinese researcher Zhang Fucheng and colleagues from the Institute of Vertebrate Paleontology and Paleoanthropology of the Chinese Academy of Sciences in Beijing used scanning electron microscopy to compare the *Confuciusornis* bone with that of modern seabirds and a middle Pleistocene galliform (Zhang et al. 1998).

Today, scanning microscopy is increasingly used in studies of dinosaur bones and teeth. Besides allowing detailed evaluations of surface topography, because scanning electron microscopy is, in general, linked to energy dispersive x-ray analysis, it permits the identification of elements that constitute dinosaur bone and teeth.

Image Analysis of Dinosaur Bone

In the late-1980s, when I began working on my PhD, I became aware of the use of computer image analyzing technology to examine forms and structures in engineering materials. I decided to apply this methodology to thin sections of bone and used the technology to assess the vascularization patterns (i.e., shape, size, and nearest neighbor assessment of the channels in the bone, which were used as a proxy for the vascular channels within the bone; Chinsamy 1993b).

Today, there are several sophisticated programs, even some free programs on the Internet, that enable researchers to do such image analysis on their personal computers. Images can be directly transferred from the microscope to the computer, and accurate measurements of various characteristics of bone microstructure can be made by using the image analysis programs.

Molecular Studies of Dinosaur Bones

Roman Pawlicki is one of the earliest researchers to search intensively for biomolecules (collagen, proteins, nucleic acids, lipids, etc.) in dinosaur bones. As early as 1966, with colleagues Korbel and Kubiak, Pawlicki described in *Nature* collagen fibers (with its typical cross-banding pattern) in the walls of vascular channels in the 80-million-year-old *Tarbosaurus* from the Gobi Desert. By 1977, Pawlicki intensively applied histochemical methods to dinosaur bone and had located lipids in the vascular channels of a *Tarbosaurus* bone using Sudan black B stain and in separate analyses showed the occurrence of mucopolysaccharides in a *Tarbosaurus* bone. Both the lipids and the mucopolysaccharides occurred in the perivascular channels, which correlates with their distribution in modern vertebrate bones (Pawlicki 1977a, 1977b).

With the advent of the polymerase chain reaction, biomolecular studies in paleontology were given a boost, even though many of the proteins and polysaccharides were not amenable to amplification. As a result, most studies of fossil biomolecules have been limited to amino acids. However, the application of modern immunological techniques of raising antibodies specific to proteins or polysaccharides had a major effect in identifying specific biomolecules in dinosaurs and other extinct vertebrates. For example, Muyzer and colleagues (1992) used antibodies raised against osteocalcin, a noncollagenous protein that is found in the bone matrix to identify osteocalcin in the bones of dinosaurs.

Mary Schweitzer is currently the most active researcher in molecular studies of dinosaurs. She has successfully applied advanced immunocytochemistry techniques to dinosaur bones (e.g., the location of heme compounds in the trabecular bone of *T. rex;* see, e.g., Schweitzer 1997). By applying high performance liquid chromatography, Schweitzer identified organic molecules (glycine and praline, which are important constituents of collagen) in an extract of *T. rex* bone (Schweitzer 1997). In another study, using both transmission and scanning electron microscopy, mass spectrometry, and x-ray photoelectron spectroscopy, Schweitzer and colleagues (1999) described the presence of beta keratin in the pedal ungual phalanx of *Rahonavis ostromi*. Since beta keratin is unique to birds and reptiles, the researchers propose that they are remnants of the keratin claw sheath that would have covered the phalanx. Schweitzer has also discovered beta keratin in the featherlike integument associated with *Shuvuuia deserti,* which is taken as evidence that the structures are indeed feathers (Schweitzer et al. 1999).

Conclusions

When a dinosaur is discovered, a host of studies of both the depositional environment and the actual remains of the animal are activated. Detailed taphonomic studies provide an assessment of how the bones came to be where they were found, while anatomical studies of the bones and cladistic analyses permit researchers to unravel the taxonomy of the dinosaur and so establish what kind of dinosaur has been discovered. Other macroscopic studies of the bones involve assessing the functional morphology and biomechanical functioning of the skeleton, and by applying the extant phylogenetic bracket approach, reliable deductions about soft tissue anatomy can be reconstructed. Microscopic studies permit deductions regarding the nature of the bone tissues preserved and involves thin sectioning the fossil bone. Procedures for preparing both fossil and modern bone are carefully explained. Electron microscopic studies of dinosaur bones have provided an assessment of the surface topography of dinosaur bone, while molecular studies document biomolecules that may have survived the fossilization process.

CHAPTER THREE

From Bone Microstructure to Biology

That bone microstructure of extinct vertebrates generally survives the fossilization process permits researchers to decipher a host of information held within the internal structure of the bone. Once a thin section of a fossil bone is obtained, it can be examined under the microscope, and information about the local growth processes and remodeling activities that affected its growth can be deciphered by comparing it with modern taxa. The molecular and cytological levels of organization of bone appear to have evolved early in the evolutionary history of vertebrates and has remained consistent throughout this time. However, at a microstructural level, the arrangement of the tissue varies in response to a number of factors such as ontogeny, phylogeny, biomechanics, and habitat (e.g., Enlow and Brown 1958). A high degree of convergence of bone tissues is recognized among various vertebrate lineages (irrespective of phylogenetic relationships) that are linked to similar adaptations (e.g., for aquatic lifestyles; Francillon-Vieillot et al. 1990). Therefore, although the phylogenetic bracket needs to be considered in trying to interpret the bone microstructure of fossil vertebrates, it is important to note that bone tissues are not constrained by phylogeny alone.

Bone Growth and Morphogenesis

As early as 1798, John Hunter recognized and formulated the fundamental concept of skeletal growth involving the two mutually related processes of outer deposition and inner resorption (Fig. 3.1A). In the case of a long bone, remodeling through bone resorption and deposition involves the longitudinal and appositional restructuring and reshaping of the bone as it grows. In 1963, Donald Enlow published his classic work titled *Principles of Bone Remodeling.* This work is still essential reading for anyone entering the field of bone microstructure because it provides the best documentation of the processes that affect how long bones grow.

Unlike most soft tissues, bone does not grow by interstitial expansion; it grows instead by a process of appositional surface deposition. Various "surfaces" are recognized, namely, periosteal (from periosteum), endosteal (from endosteum), inner surface of osteons, surfaces within resorption spaces, and so on. However, bones do not simply grow by addition of bone on the surfaces. Growth in length is achieved through endochondral growth (see the section "Femoral Growth in Length"), while growth in diameter involves both the periosteal and endosteal surfaces of the bone: the appositional increase of bone on either the periosteal or endosteal surface involves the corresponding removal of bone from the contralateral surface (a process termed *drift*), for example, as the diameter of the shaft of the bone increases by periosteal bone deposits, bone is removed from the endosteal margin (Fig. 3.1). In addition, bone undergoes complex structural remodeling in which areas of the bone are relocated to adjust for the size, shape, and axis of the bone. For example, because the metaphyses of the young long bone is wider than the shaft of the older bone that would be later derived from it, the metaphyseal region of the young bone would need to decrease in diameter to be incorporated into the shaft of the growing bone. This localized process of resizing and reshaping is termed *metaphyseal remodeling* or *metaphyseal reduction* and involves periosteal resorption of bone and endosteal deposits of bone (Enlow 1962; Fig. 3.1).

Such remodeling processes are reflected in the texture of the bone microstructure and result in a distinct stratification of the cortex. (Note that according to Enlow the term *cortical stratification* refers to the layering of different bone tissue [microstructure] types in the compacta and does not include stratification within a given bone tissue type.) Thus the presence of a particular type of bone tissue and the stratification of the compacta indicates the remodeling processes that occurred. For example, the presence of compacted coarse cancellous bone indicates that this tissue was produced during endosteal growth and was once located nearer the ends of the younger bone or encountering inner circumferential lamellae in a cross section of a bone directly indicates that it was formed in the absence of cancellous bone and is the result of metaphyseal and diaphyseal remodeling. In the same way, the presence of reversal lines indicates a change in the direction of growth.

Thus, metaphyseal reduction, reversal, cortical drift, and tubercle formation and secondary reconstruction (i.e., resorption and redeposition of bone) all contribute to

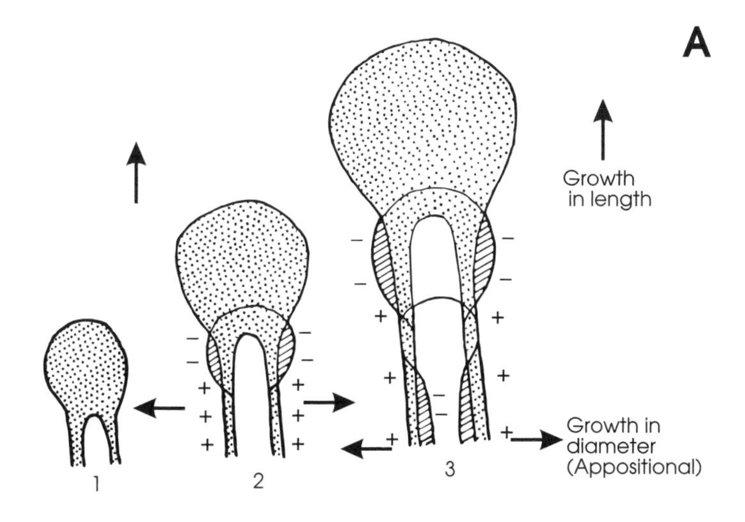

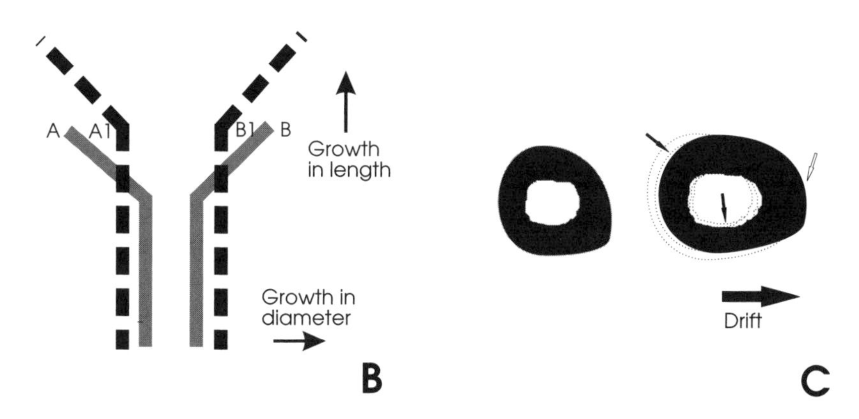

Fig. 3.1. *A*. Growth in length (vertical arrows), growth in diameter ("+" and horizontal arrows), and bone resorption ("–"). The overlay of an increasingly older bone on younger bone shows how, as a bone grows in size, old regions are relocated into different parts of the new bone, while bone resorption and addition maintains the shape of the bone (redrawn from Francillon-Vieillot et al. 1990). *B*. Metaphyseal reduction. As the bone increases in length, the metaphysis (funnel-shaped end of the bone) of the younger bone has to be reduced in diameter so that it is incorporated into the diaphysis of the older bone. Basically, region AB in the younger bone would correspond to region A1B1 in the older bone (redrawn from Enlow 1969). *C*. As the bone grows, the diaphysis of the bone undergoes remodeling changes to adjust for alteration in the shape and axis of the bone. Here, the cortex "drifts" in the direction indicated by the arrow by a combination of bone deposition (open arrow) and bone resorption (filled arrows) on either the periosteal or endosteal surfaces.

the formation of a stratified cortex. For this reason, bone microstructure in different parts of a skeleton, as well as in different parts of the same bone, can be quite different. The more intricate the bone morphology, the more the bone needs to involve remodeling processes (such as drift) to accommodate changes. Thus, because of the tubular shape of long bones, it is recognized that their bone microstructure, particularly in the midshaft or neutral region, best preserves the integrity of the depositional history of the bone during growth (although it must be noted that this is rarely a complete account because much of the bone formed in early ontogeny is obliterated as a result of growth and remodeling changes that occurred during later growth).

Femoral Growth in Length

In the developing embryo, femora are first formed as a cartilagenous precursor, with an external fibrous tissue called the *perichondrium* that consists of chondroblasts that form the cartilage (Fig. 3.2). As with all endochondral bones, the shape of the bone appears to be determined by the cartilagenous prototype that appears to be controlled by genetic and developmental factors (Smith and Hall 1991). Although stress exerted

Fig. 3.2. (*opposite*) *A.* How a bone grows. *1.* Cartilage precursor of the long bone. *2.* Development of a bony collar around the shaft of the bone (diaphysis). *3.* The calcification of cartilage in the diaphysis is followed by an invasion of blood vessels, and the zone of calcification drifts toward the metaphysis and continues to advance toward the epiphysis and form what is often referred to as the *growth plate. 4.* Here, the calcification of the cartilage continues, and the bone grows in length. This layer is usually overlain by uncalcified cartilage and articular cartilage, but these are not preserved in fossil bones. Thus, what we see as terminal ends of a dinosaur bone are really the ends minus the articular and uncalcified cartilage. *B.* Stages of endochondral ossification in a long bone. *C.* Different strategies for growing in length in crocodiles, mammals, and birds. In the crocodile, growth in length occurs under the overlying articular cartilage and is able to continue indefinitely. Mammals, however, develop secondary centers of ossification early in development (round about stage 3), which also ossifies in both an outward (i.e., toward the articular ends of the bone) and inward (toward the growth plate). A network of cartilage canals are located in the secondary ossification center and also penetrates the growth plate and can join with extending marrow processes below. When ossification of the secondary center meets with that of the growth plate, growth in length ceases. Birds also have determinate growth, but they do not develop secondary centers of ossification. However, the growth plate in birds is penetrated by cartilage canals, which penetrate and meet with the extending marrow processes from the diaphysis, which seems to facilitate endochondral bone formation. Arrows indicate the direction of endochondral growth and (a) indicates articular cartilage. Diagrams are based on illustrations from Haines (1942) and Reid (1997).

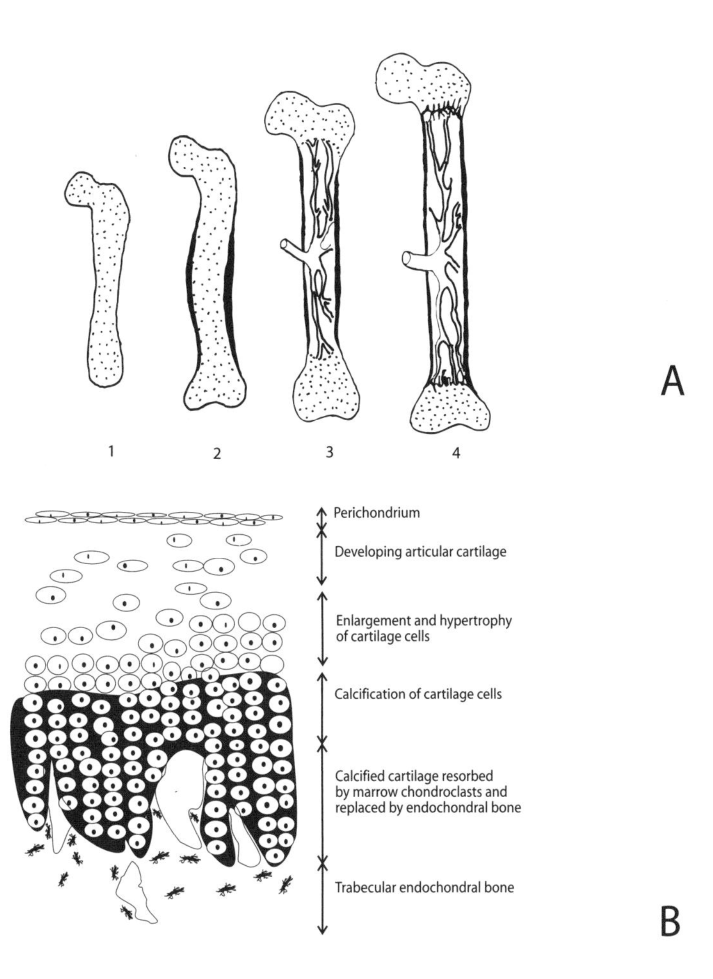

1
2
3
4
A
Perichondrium
Developing articular cartilage
Enlargement and hypertrophy of cartilage cells
Calcification of cartilage cells
Calcified cartilage resorbed by marrow chondroclasts and replaced by endochondral bone
Trabecular endochondral bone
B

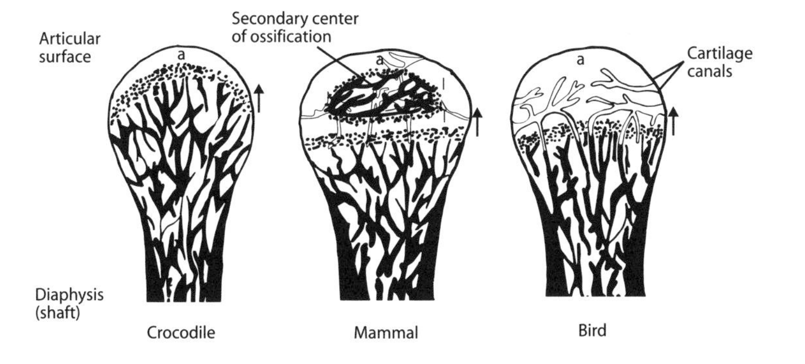

Articular surface
Secondary center of ossification
Cartilage canals
a
a
a
Diaphysis (shaft)
Crocodile
Mammal
Bird

C

on the developing bone does not affect its shape, it does affect the internal organization of the bone (Thomason 1995).

Initially growth involves the addition of more cartilage, but later bone-forming cells, called *osteoblasts,* appear in the perichondrium and begin to deposit bone over the cartilage surface. About the same time, a second process begins that results in the calcification of the central part of the shaft and in the invasion of cartilage-absorbing cells, called *chondroclasts,* into this area. As the chondroclasts resorb the cartilage, osteoblasts begin to line the cavities with bone. These processes continue in the growing bone and eventually spread along the shaft of the bone to the proximal and distal parts but not as far as the actual terminal surfaces, which remain cartilagenous longer and become the articular cartilages. Thus growth in thickness continues by the accretion of more periosteal bone along the sides, while growth in length continues through the formation of new cartilage at the ends and its subsequent replacement by endochondral bone (Fig. 3.2).

In living animals, uncalcified cartilage, as well as a zone where the cartilage cells become hypertrophied and then packed in columns, overlay this region. Note that these regions of cartilagenous growth are not preserved in fossil bones and thus the seemingly articular ends of the dinosaur bone actually represent the region between calcified and uncalcified cartilage (Fig. 3.2B). This type of longitudinal growth, where replacement of cartilage by endochondral bone occurs under the articular cartilage, is seen in turtles and crocodiles and differs from the manner in which longitudinal growth occurs in mammals, lepidosaurs, and some amphibians. In the latter tetrapods, growth in length occurs at an internal growth plate, which extends transversely between the calcified or ossified epiphysis and the expanding metaphysis (Fig. 3.2C). In this case, the cartilage near the articular end is not involved in the longitudinal growth. In mammals, on maturity, the ossified epiphyses fuse with the metaphyses and thus prevent further growth in length. This feature makes juveniles and adult mammal readily discernible. Birds also have determinate growth, but they do not have secondary ossification centers (although there are reports of possible secondary centers in the proximal end of the tibia in some birds). Birds, however, unlike crocodilians, do not continue to grow in length indefinitely. Interestingly, like mammals and other animals that have determinate growth (e.g., many lizards), articular ends of bird bones have cartilage canals that penetrate the growth plate region (Fig. 3.2C).

There is currently confusion in the literature regarding the use of the term *epiphysis.* To many, this term refers to the development of the secondary ossification center (particularly among mammals) in the cartilage overlying the growth plate. However, these days, the term *epiphysis* follows that of Haines (1942), where it applies to the whole of the cartilage mass at the end of a bone, irrespective of whether a secondary ossification is present. Thus, *epiphyses* refers to the cartilaginous ends of long bones of mammals, birds, and reptiles alike and secondary ossification centers that develop in the cartilage are referred to as such.

Dennis Carter and colleagues (1998) propose that the development of secondary ossification centers depends on the interaction between mechanical epigenetic factors and genotypic factors that determine the rate of perichondral (periosteal) and endochondral ossification. Using finite element computer models, they found that when perichondral ossification significantly precedes endochondral ossification the resulting mechanical stresses in the developing cartilage anlagen inhibits the formation of secondary ossification centers (Carter et al. 1998).

Rate of Bone Formation

In 1947, the Italian researcher Amprino proposed that the differences in bone tissue types are a reflection of variation in bone depositional rates, which not only determines the volume of bone laid down but also influences the organization of the fibrillar matrix and the type of tissue that results. Thus, in life, bone is a living tissue that directly records its rate of formation.

In a fossil bone, although the organic components (such as collagen, osteocytes, and blood vessels) are generally not preserved, the arrangement of the crystals of the bone mineral, which followed the organization of the collagen matrix in life, can be assessed (see chapter 1; Fig. 1.2). This permits deductions about the type of bone tissue that was present and allows inferences about the qualitative rate at which bone formed in the extinct animal. As discussed in chapter 1, fossil bones preserve various other microstructural details, such as the lacunae where the bone cells (i.e., the osteocytes) were once located, the fine canaliculi, which facilitated communication between neighboring bone cells and the cavities through which blood vessels once traversed (see Fig. 1.3).

According to Amprino's hypothesis, a slow rate of bone formation results in a lamellar bone matrix, which characteristically has an ordered arrangement of collagen fibers and osteocytes that tend to be more flattened. However, rapid bone formation results in fibrolamellar bone tissue in which the collagen fibers are more haphazardly arranged and form a fibrous, or woven, bone matrix. This type of bone matrix also contains randomly distributed globular osteocytes. During the rapid formation of this woven matrix, numerous blood vessels become entrapped and later more slowly formed lamellar bone is formed around each blood vessel producing primary osteons. Lamellar and fibrolamellar bones are extremes of slow and fast rate of bone formation; intermediate rates also occur, as in the parallel-fibered type of bone tissue, which has an intermediate arrangement of the collagen fibrils and lamellated bone that shows small-scale cyclical variations in fibrillar and mineral microstructure (Fig. 3.3).

Fig. 3.3. Different types of bone tissue based on fibrillar orientation. *A.* Fibrolamellar bone in the humerus of *Archaeornithomimus* from China. *B.* Lamellar bone with lines of arrested growth in *Timimus,* a polar dinosaur from Dinosaur Cove, Australia. *C.* Endosteally lamellated bone tissue in *Timimus.*

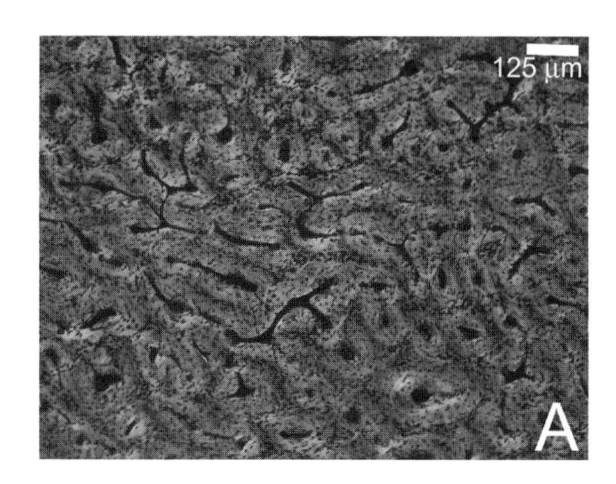

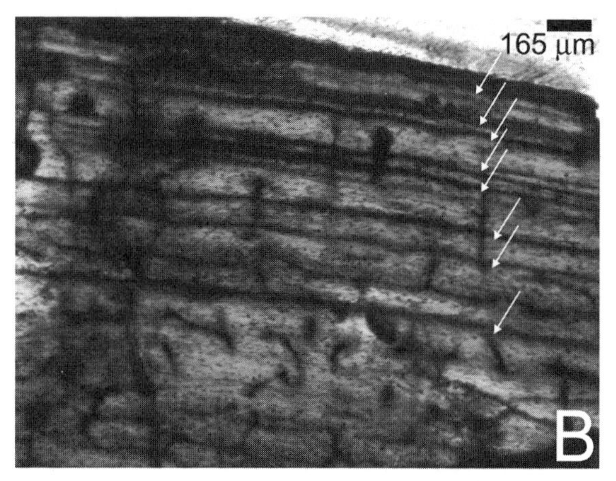

Amprino's work pertains to a qualitative description of the rate at which bone forms. To obtain quantitative information about rate of bone formation and tissue types that result, several experimental studies using fluorescent labeling have been conducted. The most widely used fluorochrome stains are tetracycline drugs that are either administered with food or by injection. The fluorochrome stains are very effective since they bind to the newly formed bone, and when the thin section is examined under ultraviolet light, the stain in the bone fluoresces. Thus, one can determine where and how much bone was deposited (see Plate 1). Experimentally, different color fluorochrome stains can be used to demarcate duration of differing periods of growth and are useful in quantifying bone depositional rates.

Comparative studies of modern animals have shown that different rates of bone deposition result in different bone types. In a comparison of published data on bone depositional rates of several vertebrates, de Ricqlès et al. (1991) found that the formation of woven-fibered bone often exceeds 40 microns per day, but can also form at 5 microns per day, while parallel-fibered bone is generally formed between 0.10 and 0.5 microns per day and compact lamellar bone results at rates that do not exceed 30 microns per day but can also occur from bone depositional rates as low as 0.04 microns per day. Thus, there is substantial overlap in the rates at which different bone tissues form.

A few years ago, Jacques Castanet of Paris VII Université and colleagues (1996) showed that different bones in a skeleton have different depositional rates and that such rates vary during ontogeny (e.g., in the mallard, he found that early humeral growth approached 40 microns per day, while at 7 weeks, humeral growth decreased to 24 microns per day). These researchers found that in the mallard, laminar and subplexiform bone tissues were formed when bone depositional rates exceeded 15 microns per day but that between 5 and 15 microns per day plexiform and reticular tissue types resulted, while a rate below 5 microns per day caused a substantial decrease in the density of blood vessels. Bone depositional rates of less than 1 micron per day, resulted in poorly vascularized parallel-fibered bone types. Subsequent to this study, Castanet and colleagues (2000) analyzed growth in young ostriches and emus and found that laminar and reticular bone formed at rates as low as 2.9 microns per day and as high as 80 microns per day in emus, while in the ostriches, rates of 2–64.9 microns per day resulted in similar tissue types.

In a more recent study on hatchling Japanese quail (*Cortunix japonica*), neontologist Matthias Starck and I (2002) documented bone depositional rates of 10–50 microns per day for fibrolamellar bone tissue, with mainly longitudinally oriented channels overlapping with the range of bone depositional rates that Castanet and colleagues (2000) found for the different tissue types. The results of our experiments further demonstrated that different bones in an individual grew at different rates (i.e., the midshaft region of the femur and humerus experienced a faster rate of bone deposition than the radius, the ulna, or the tarsometarsus), although the radius grew at the slowest rate (Starck and Chinsamy 2002). From our study, it is obvious that substantial variation in bone deposition exists among vertebrates, which is affected by ontogeny, location in the skeleton, and local growth conditions (Plate 1). Thus, Amprino's hypothesis that different bone tissue types are a reflection of depositional rates still holds true qualitatively. However, quantitative data from fluorochrome labeling suggest that the different tissue types are not formed at unambiguous rates (i.e., 10 microns per day in the mallard results in plexiform to reticular types), while in the Japanese quail, this rate of bone formation (osteogenesis) results in laminar bone tissue.

Cyclical versus Noncyclical Growth

The overall structure of primary compact bone provides a direct assessment of whether bone deposition was continuous or interrupted (Fig. 3.4). A cyclical rate of bone deposition is evident in the compacta as distinct growth rings or alternating bands of tissue types. This type of tissue is termed *zonal bone* and the resulting bands of tissue are the *zones* and *annuli*. The zones tend to be more vascularized and represent periods of fast growth, while the poorly vascularized annuli, which usually tends to be lamellar bone, signal periods of slower growth. Pauses in osteogenesis are marked by distinct lines of arrested growth (LAGs), which may precede or follow annuli (Figs. 3.4, 3.5).

Growth Rings in Reptiles

Among extant reptiles, it is generally found that a single zone and annulus is formed per year (Figs. 3.4, 3.5A, 3.5B). Experiments with fluorochrome dyes, such as tetracycline and xylenol orange, have verified that this pattern of bone deposition is seasonally related, with zones forming during the warmer months and annuli resulting during unfavorable periods. Thus, by counting the number of zones and annuli, the age of the animal can be obtained—a methodology termed *skeletochronology*. Skeletochronology has been applied to crocodiles (e.g., Hutton 1986) and various extant lepidosaurs (lizards and snakes; e.g., Castanet and Smirina 1990; Castanet and Baez 1991; Castanet et al. 1993). Double or accessory lines can sometimes occur in the bone tissue (e.g., biannual cycles of activity in a newt, *Triturus marmoratus,* living at high altitudes have been reported by Caetano 1990). However, there are no reports of only one LAG formed every two years, which supports the deductions that LAGs in fos-

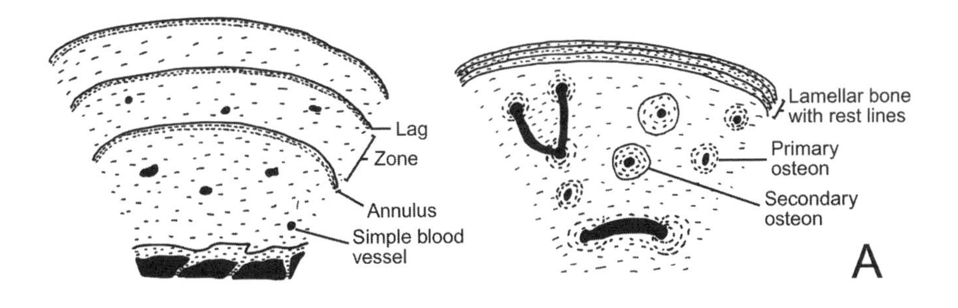

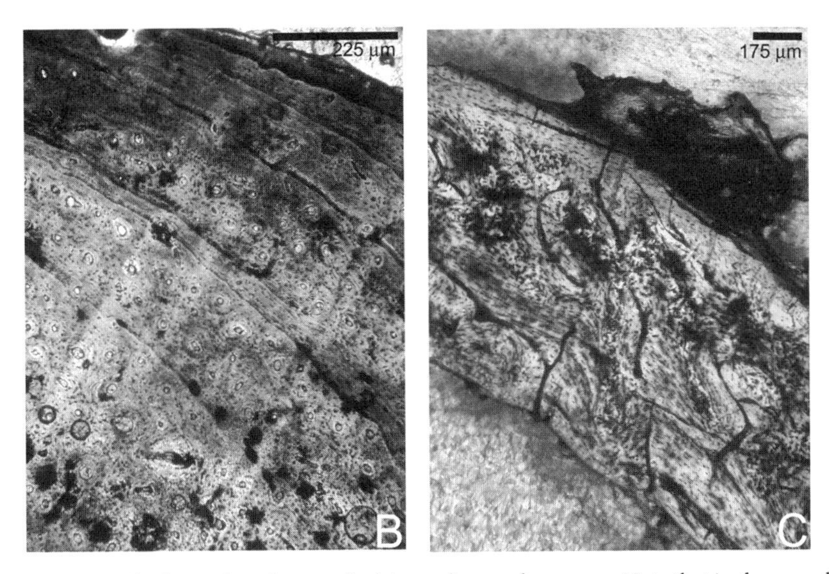

Fig. 3.4. *A.* Cyclical (zonal) and noncyclical (azonal) growth patterns. Note that in the azonal bone, once somatic maturity has been attained, slow deposits of lamellar bone occurs. Rest lines are sometimes located in this outermost part of the compacta. *B.* Cyclical bone in a nonmammalian therapsid. *C.* Noncyclical growth in a rabbit femur. Note the change from a fibrolamellar bone tissue to a lamellar type of tissue near the perosteal surface of the bone.

sil vertebrates are more than likely annual. Seasonality (i.e., hot-cold in temperate regions and wet-dry in tropical regions) appears to be the main factor responsible for the cyclical growth pattern recorded in the bones of reptiles (e.g., Peabody 1961). However, the occurrence of LAGs in bones of reptiles and amphibians living in tropical or aseasonal environments (e.g., Castanet and Baez 1991; Chinsamy et al. 1995) suggests that the cyclical variations in osteogenesis and growth are the result of an inherent (genetic) rhythm that becomes synchronized with and reinforced by

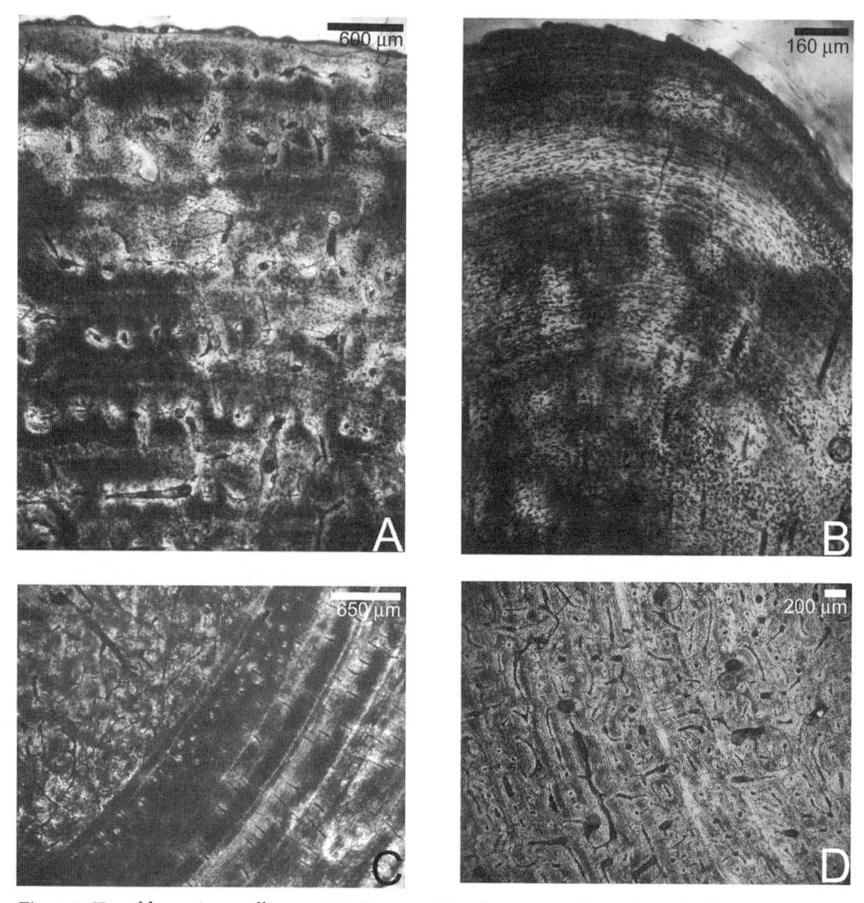

Fig. 3.5. Zonal bone in an alligator (*A*), *Varanus* (*B*), a kangaroo (*C*), and a polar bear (*D*).

seasonal cycles (e.g., Patnaik and Behera 1981; Castanet et al. 1993). That the spacing between consecutive growth rings changes (becomes closer) after sexual maturity has proved to be useful in assessing age at which sexual maturity occurred. In our study of *Angolosaurus skoogi,* a cordylid lizard from Namibia (Chinsamy et al. 1995), we were able to develop growth curves for males and females of the taxa (which were shown to be different), and on the basis of the spacing of consecutive LAG deduced that sexual maturity occurred between the second and third year of life. We were also able to verify that *Angolosaurus* is a particularly long-lived lizard with up to ten growth rings recorded in the bone of a female.

Studies in skeletochronology have provided valuable insight into studies of population biology and evolutionary processes of various extant amphibians and reptiles (Castanet et al. 1993), and although its potential for use in fossil reptiles (to assess life history parameters, as well as to deduce paleoecology) has long been recognized (Peabody 1961), it has still not yet been fully explored.

Growth Rings in Mammals

Annual growth marks are relatively well recognized in epidermally derived structures, such as the claws of seals, the baleens of whales, the horn sheaths of mountain sheep, but are rarely encountered in bones of mammals and birds (Peabody 1961).

Studies of growth marks in mammalian bones were initiated and first recognized by the late Sergei Evgenievitch Kleinenberg of the Severtsov Institute of Animal Morphology in Russia and his student at the time Galina Klevezal and have contributed immensely to understanding growth marks in mammalian bones. The results of their study were first published in Russian in 1967 and fortunately for the wider scientific community translated into English in 1969 by the Israel Program for Scientific Translations. Klevezal and Kleinenberg (1969) found that, unlike zonal bone that formed in the compact bone of reptiles, among mammals, the "recording" part of bone is the periosteal zone (i.e., the outermost part of the cortical bone, which is generally formed once the overall body growth rate has slowed down; Fig. 3.4A). In juvenile mammals, bone is generally a highly vascularized reticular or plexiform type of bone tissue without any interruptions in the rate of growth. However, later in ontogeny, many mammals begin to deposit dense bone periosteally and can form rest lines in this region. Therefore, among mammals, only the periosteal zone of the compact bone is a recording structure of growth. As Klevezal, now of the Institute of Developmental Biology of the Russian Academy of Sciences, explains, because they are not formed from the onset and only later in ontogeny, a count of the rest lines will not give a true account of age and needs a correction factor to adjust for the time during which no growth rings were formed (Klevezal 1996). This correction factor varies for different animals because in some mammals periosteal bone with resting lines form from the first year of life onward, while in others, it does not form for several years. Klevezal has found that among sika deer periosteal bone with rest lines is deposited only after two to three years, while in the polar fox and mink, they form in the first year. A further complication is that in some mammals periosteal bone with rest lines are rare throughout ontogeny (e.g., in baleen whales except for the dense auditory bones, most skeletal elements do not form rest lines at all). In fact Klevezal suggests that for fast-growing mammals one has to select carefully bones that demonstrate negative allometry (i.e., bones that have a slower growth rate then the rest of the skeleton). Thus, when the growth rate is high in mammals, periosteal bone cannot be considered a recording structure; however, when there is slow incremental growth, then rest lines are formed in the periosteal bone.

Klevezal notes that the best structures for recording growth in mammals are mainly cementum and dentine of teeth and only on occasion is bone useful for such analyses. *Cementum* is regarded as the best for skeletochronology since it forms during the entire life of the animal and it is unaffected by remodeling and reconstruction. *Dentine* has a more limited potential because it forms only until the pulp cavity is filled, and it is affected by seasonal, intraseasonal, and daily changes in growth rate and calcification. Periosteal bone in mammals is generally not considered good for age determination since it persists for a limited amount of time (Klevezal 1996).

In her detailed studies of recording structures, Klevezal has found that the growth rate of the particular structure can change in concert with the growth rate of the whole organism. She has suggested that seasonal changes of growth intensity of an animal as a whole determine the formation of growth layers. Thus, a seasonal change of growth rate of a given recording structure reflects the seasonal growth rhythm of an individual.

Until relatively recently, it was believed that growth rings occurred in only hibernating mammals and mammals from cold regions. However, today it is recognized that even mammals from temperate regions can form growth rings in recording structures of cementum, dentine, and bone. Mammals in the Tropics do not show distinct layers in recording structures because there is no seasonal factor, though Klevezal (1996) suggests that perhaps changes in humidity, not temperature, may play a seasonal factor in growth.

In the classical 1969 study, Klevezal and Kleinenberg found growth lines occurring in the bones of animals from cold regions and marine habitats, as well as in mandibles of small lagomorphs and rodents. More recently, Guy Leahy (1991) reported growth lines in the bones of a kangaroo. I have observed such lines in kangaroo bones (Fig. 3.5C) and have also documented growth lines in a femur of a wild polar bear from the Arctic Circle (Fig. 3.5D). It is therefore apparent that growth lines are not restricted to reptiles. More research is needed to ascertain what exactly triggers the switch from uninterrupted fibrolamellar bone to an interrupted type in mammals. Temperature appears to be an important factor but perhaps diet or, as Klevezal (1996) suggests, maybe even humidity may affect the growth rate of the animal. These are compelling questions that future research needs to assess.

Given the restrictions discussed, it is evident that growth marks in the bones of reptiles and mammals show a periodicity that appears to be correlated to the growth dynamics of the animal. Thus, they can be effectively used for ascertaining growth patterns, morphogenesis, and individual age.

Noncyclical Bone Formation

A sustained, uninterrupted bone formation results in the lack of zonation in the compacta. The tissue comprising the compacta can be entirely of fibrolamellar bone or can grade into a parallel-fibered or lamellar type of tissue. As mentioned previously, fibrolamellar bone is common among mammals and birds and results from a rapid rate of bone formation. However, this uninterrupted, azonal type of bone formation is not restricted to mammals and birds only. Similar tissues have been reported in *Diictodon,* a dicynodont, where such tissues formed for most of the animal's life (Chinsamy and Rubidge 1993; Ray and Chinsamy 2004), and in the nonmammalian cynodonts, *Cynognathus,* and *Tritylodon* (Botha 2001), as well as in some dinosaurs, for example, *Dryosaurus* (Chinsamy et al. 1995b), *Centrosaurus* (Tumarkin-Deratzian 2003), and pterosaurs (de Ricqlès et al. 2000). The presence of uninterrupted fibrolamellar bone tissue indicates that bone deposition was rapid and noncyclical.

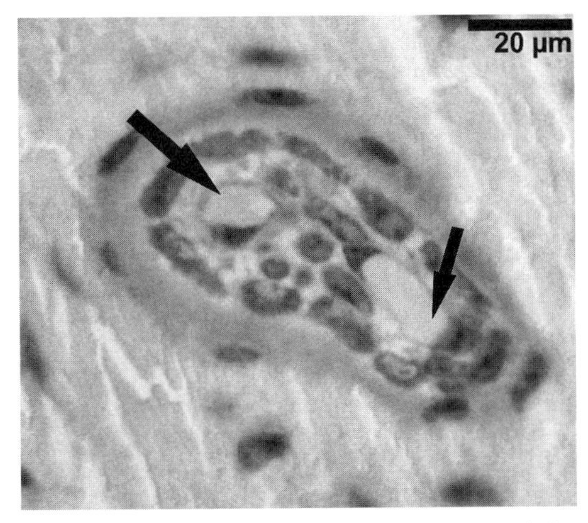

Fig. 3.6. Two blood vessels (arrows) inside a channel of a modern bird (*Coturnix japonica*). Note that the shape of the channel has no bearing on actual shape of blood vessels therein. Image courtesy of Matthias Starck.

Extent of Vascularization

Inspecting a thin section of a fossil bone under a microscope can reveal the amount and the type of "vascularization" that were present in the once-living bone. In the past, it was generally assumed that the area of the spaces was equivalent to the area occupied by blood vessels. However, in Starck and Chinsamy (2002), we found that in juvenile Japanese quail only about 20% of the area of the channel is occupied by blood vessels and that one or more blood capillaries are located within the spaces (Fig. 3.6). (With maturity, the space in the channel would be occupied by centripetal-lamellar bone to form primary osteons, but even then, blood vessels would not fill the channel entirely.) Nerves and connective tissue occupy the rest of the space in the channel. Another important finding was that the shape of the cavity bore no real resemblance to the actual shape of the blood vessels therein.

Depending on the type of vertebrate and the rate at which the bone was formed, avascular or vascular bone can occur. As the name suggests, avascular bone is devoid

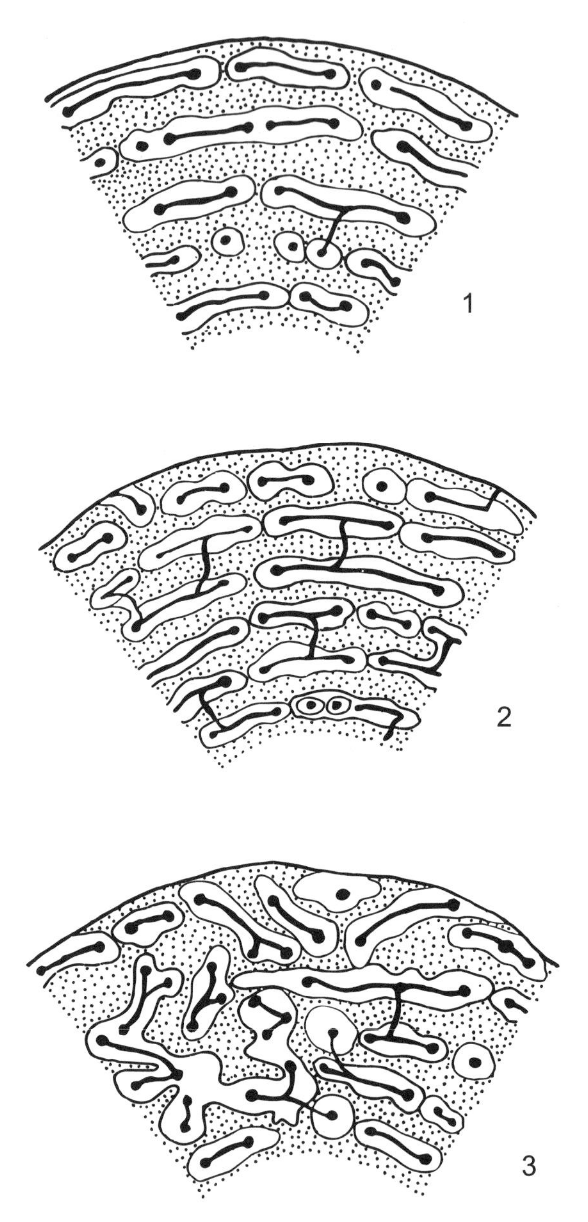

Fig. 3.7. Different types of fibrolamellar bone tissue. This differentiation assumed that the channel shape reflected the shape of the blood vessels. Even though we know today that this is not the case, the categorizations can be used as long as one is clear that they refer to the orientation of the channels (and not the blood vessels) in the bone. 1. Laminar bone shows a regular succession of channels parallel to the external surface of the bone. 2. Plexiform bone is similar to the arrangement in 1 but also has additional radial connections. 3. Reticular bone has irregular channels with no particular arrangement. Redrawn from de Ricqlès (1974).

of a vascular network. Here the osteocytes tend to develop a more extensive system of intercommunicating canaliculi, which probably compensates for the lack of blood vessels. Vascular bone, however, consists of a network of blood vessels that can be dense or sparse. The orientation of the vascular canals was thought to depend on the form of the spaces that were initially enclosed. Armand de Ricqlès (1974) equated the channels in bone with vascular channels and used these to define topological categories of tissues within the fibrolamellar complex (Fig. 3.7). Thus, in laminar bone, the blood vessels are organized in concentric layers around the cortex, while in plexiform bone, there is a similar arrangement with, in addition, radial connections between vascular canals. In reticular bone, the connections have random orientations. A radial arrangement of channels can occur in vertebrates, although this is uncommon in dinosaurs. This terminology is problematic because we now know that the shape of the channels has no real bearing on the shape of the blood vessels. If these categories described by de Ricqlès are used, one needs to be clear that the description is based on the arrangement of the channels and not on the actual blood vessels. This point is exemplified because, in some of the longitudinally oriented channels, Starck and I found circumferentially oriented blood capillaries (Starck and Chinsamy 2002; see Fig. 3.6).

Simple blood vessels, primary osteons and secondary osteons, are relatively easily recognized in a thin section of bone. Simple blood vessels are ones in which there is no change in the bone microstructure around the vascular channel (i.e., the blood vessels are simply incorporated in the appositional growth of the bone; Fig. 3.8A). Primary osteons result when bone is formed rapidly (e.g., when fibrolamellar bone is deposited) and spaces are left around the blood vessels (see Fig. 3.9A). Later on, these spaces are infilled by the centripetal deposition of lamellar tissue around the blood vessels (Fig. 3.9B; Plate 2). Thus, primary osteons are distinguishable by a bony ring of lamellar tissue around each blood vessel (Figs. 3.3A, 3.8B). The presence of primary osteons is indicative of a rapid rate of bone deposition, whereas simple blood vessels occur when bone tissue is more slowly formed. Secondary osteons do not reflect the rate at which the bone was formed but are the result of a process of secondary reconstruction (Fig. 3.8B). This involves the removal of bone around a primary vascular canal, followed by subsequent redeposition of concentrically arranged lamellar bone in the erosion cavity. Secondary osteons are easily distinguishable from primary osteons by the presence of a cement line or reversal line, which marks the furthest extent of bone removal. Successive generations of secondary osteons can form in the same location and will result in dense Haversian bone tissue in which the interstitial bone between the secondary osteons is also secondary in origin (Fig. 3.8C).

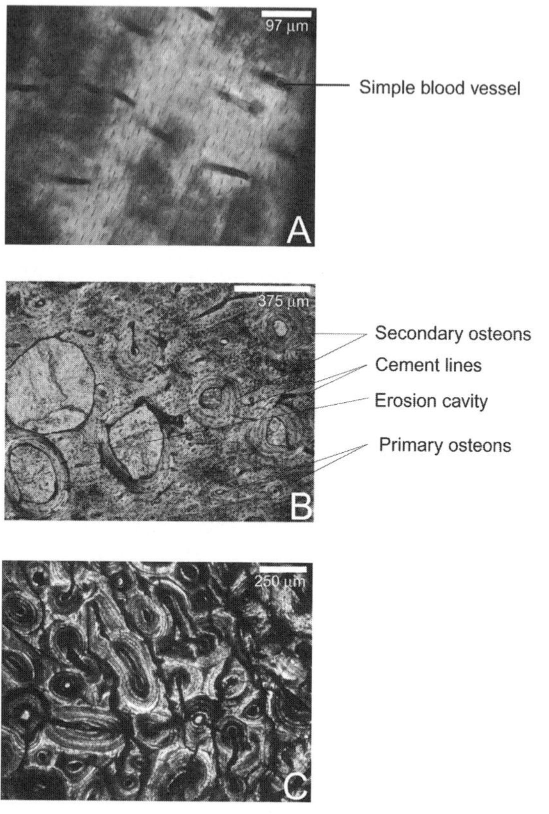

Fig. 3.8. Different types of vascularization. *A.* Simple blood vessels in a kangaroo femur. *B.* Primary and secondary osteons in *Kentrosaurus,* a stegosaur from the Tendaguru deposits in Tanzania. *C.* Dense Haversian bone in a sauropod bone from Niger.

Bone biologist Lacroix (1971) determined that the time taken to form a secondary osteon could be assessed by addition of the time taken for resorption and that for redeposition of the lamellar bone around the resorption cavity. I have used Lacroix's formula to estimate a minimum age of eighty days (with a deposition rate of less than 1 micron per day) for a young *Dryosaurus* that had just begun to undergo Haversian reconstruction in its femur (Chinsamy 1995b). It is also possible to multiply the number of generations of osteons by the number of days it takes to form a single osteon to estimate the minimum age of an animal. However, exercise caution because Haversian reconstruction can occur at different stages of ontogeny, and because secondary osteons are not formed continuously, there is no way of determining the time lapse between each generation of osteons. Furthermore, given the variation documented for periosteal bone, it is possible that there may be variation in the rate at which the centripetal lamellar bone is deposited.

Numerous suggestions have been proposed regarding the biological significance of secondary osteon formation. These have included metabolic starvation resulting in damage to the vascular system, physical necrosis of osteocytes, as well as biomechanical adaptation to physical strain. One of the arguments that has received the most support is that secondary osteons are the result of phosphocalcic metabolism. This is the argument that Robert Bakker (e.g., 1972, 1986) used to propose that the dense Haversian bone in dinosaur bone implies high metabolic rates and hence endothermy in dinosaurs. It has, however, now become increasingly clear that secondary osteons are not reliable indicators of metabolic rates. Secondary osteons are known to occur in bones of turtles, crocodiles, and tortoises but are absent in small mammals such as rodents and bats, as well as in passerine birds, which are highly active.

Ontogenetic Status

In general, surface texture, articular surfaces, and open or closed sutures of an animal's skeleton can indicate its ontogenetic status. However, on occasion, one is faced with a specimen whose ontogenetic status is not readily discernible on purely morphological grounds. In such circumstances, the nature of its bone microstructure can provide substantive evidence regarding its ontogenetic status (Fig. 3.9; Plate 2).

Juveniles typically have a bone tissue that is largely cancellous in structure. This is a direct result of the rapid rate at which bone is being deposited, which, among many vertebrates, tends to result in a woven bone tissue with haphazardly arranged, large, globular osteocytes. The woven bone provides two to three times as much surface area for the deposition of the bone matrix than does lamellar bone formation and permits more bone to be deposited in a given time. With increasing age, the porous spaces are infilled with a lamellar type of bone tissue and primary osteons result. The resulting woven bone with primary osteons increases the strength of the developing bone because it has a higher Young's modulus than woven bone without primary osteons. (In more slowly growing animals, e.g., in many lizards, the juvenile tissue consists of parallel-fibered tissue with simple blood vessels.) With maturity, the animal generally switches from an earlier, more rapidly formed type of bone tissue to a more slowly formed tissue. For example, an earlier fibrolamellar bone tissue pattern may change to a lamellar, parallel-fibered or lamellated type.

Bone formed late in life typically shows a slow rate of accretion. Depending on the nature of the tissue deposited during late growth, it is possible to deduce whether the animal stopped growing (i.e., had a determinate growth, as in mammals and birds) or whether it continued growing (albeit at a slower rate) throughout its life (i.e., had an indeterminate growth strategy, as in many reptiles). In mammals and birds that experience

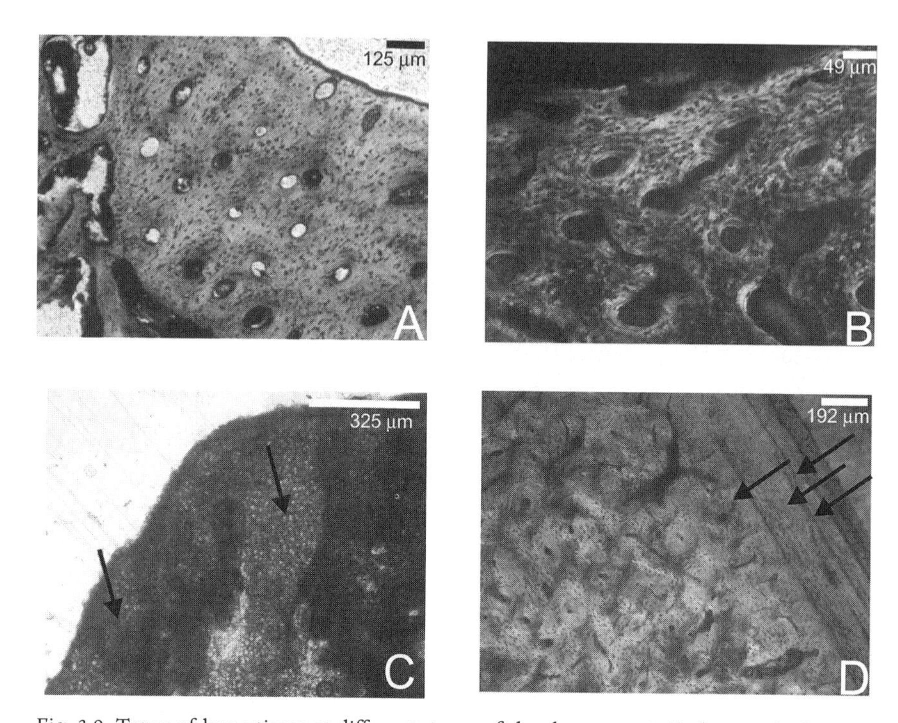

Fig. 3.9. Types of bone tissue at different stages of development. *A.* Embryonic hadrosaur showing early stages of fibrolamellar bone. Note the haphazard arrangement of the osteocytes and the open channels within the bone. Later in development, the latter will become infilled to form primary osteons. *B.* Juvenile *Nqwebasaurus* from the Algoa Basin in South Africa showing the development of primary osteons. *C.* Longitudinal section of femur of *Nqwebasaurus* showing large amount of calcified cartilage (arrows). *D.* Determinate growth reflected in a dog femur by the presence of lamellar bone (=outer circumferential layer) and rest lines (arrows) in the peripheral regions of the bone.

determinate growth, terminal growth is indicated by a change from fibrolamellar bone to a more slowly formed layer of bone in which a series of closely spaced peripheral rest lines can occur (Figs. 3.4A, 3.9D, 3.11). This outer circumferential layer is also referred to as the *circumferential lamellae* (Ham 1953), *outer circumferential lamellae* (Enlow and Brown 1957), and later as the *external fundamental system* (EFS). *Outer circumferential layer* is preferred (i.e., substitution of Ham and Enlow and Brown's "lamellae" with "layer") because this bone can comprise either lamellar or lamellated bone (Fig. 3.11).

The extent of calcified cartilage in the longitudinal sections of the distal and proximal regions of long bones will reasonably assess the state of growth in bone length. Young individuals typically have abundant calcified cartilage preserved in the distal and proximal regions of the bone (Fig. 3.9C; Plate 3).

Sexing Bones

Whether a fossilized animal is a male or female is virtually impossible to determine from only the skeletal remains. Some progress in this regard has been made with paleoanthropological material, particularly hominids, where the pelvic girdle and jaws provide good evidence for sexual dimorphism. However, for the majority of extinct animals, we are clueless about their gender. The extraordinary preservation of oviducts in *Sinosauropteryx* is the only direct evidence of a female dinosaur. Some studies of large, single taxa samples have identified robust and gracile forms, which may indicate different sexes (e.g., in *Syntarsus*). In fact, in 1990 I interpreted the abundance of perimedullary erosion cavities in the robust form of *Syntarsus* as linked to egg laying in females. However, on hindsight I feel that given the small sample size this deduction is not well substantiated.

Soon after the adult *Oviraptor* was found with a nest of eggs (Norell et al. 1995), and speculations were made that the gender of the parent could be assessed through bone microstructure (Martill et al. 1996), Paul Barrett, now of Oxford University, and I wrote about the difficulty in making such deductions from dinosaur bones (Chinsamy and Barrett 1997). Although medullary bone is known to occur in the medullary cavity of ovulating birds where it serves as a mineral store for the formation of the eggshell during egg laying, it has not been recognized in the bones of any dinosaur (Fig. 3.10). Experimental studies on modern reptiles, including crocodiles, alligators, turtles, and lizards have also shown no evidence of medullary bone forming preovulation.

Biomechanical Adaptations

In attempting to understand the biomechanical functioning of an animal, most paleontologists consider only the external morphology of fossilized bone. Yet, the internal architecture of the bone can potentially provide a host of information about the mechanical environment of the bone (i.e., the forces that may have acted on it during life). According to Dennis Carter of Stanford University and colleagues (1998), mechanical loading caused by muscle contraction, which begins during embryogenesis, is one of the most important epigenetic factors that affect the developing skeleton. During the life of an animal, as a result of normal daily activities such as walking and running, complex forces act on the skeleton through muscular activity, joints, ligaments, and tendons, which cause stresses and strains in the bone. The mechanical response of the bone to these forces is dependent on the magnitude and the direction of the force (Carter et al. 1978), as well as on its geometry and material properties (e.g., Martin and Burr 1989). As a result, different bones and parts of bones have different mechanical responses (i.e., variable

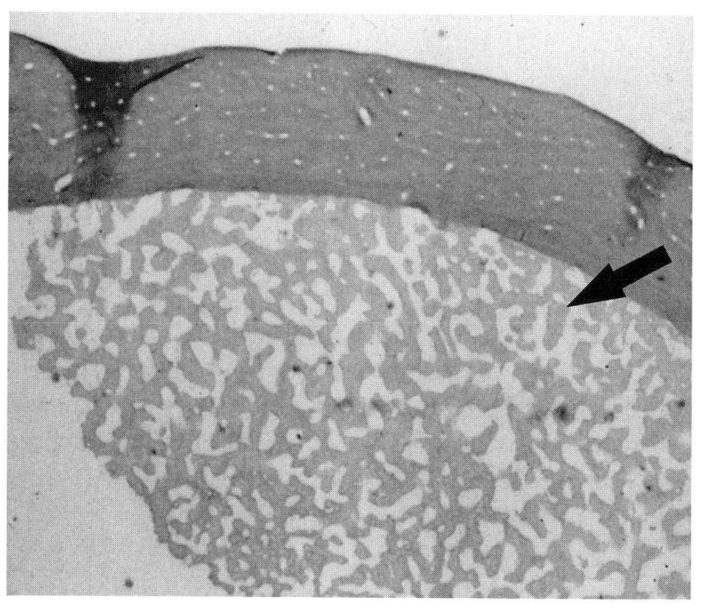

Fig. 3.10. Medullary bone (arrow) in the medullary cavity of a femur of an ovulating broad-winged hawk (*Buteo platypterus*).

strength and stiffness) because of differences in local microstructure (i.e., type of bone tissue, porosity, and mineralization) and the bone's location in the skeleton.

Bone is a composite material, and the mechanical properties of the bone are different from that of its individual components (such as secondary osteons, interstitial bone, and collagen-bone mineral). That bone can be described as a composite material at various levels further complicates the mechanical properties of bone (e.g., Carter and Spengler 1978; Currey 1991; Thomason 1995). For example, secondary osteons are embedded in interstitial bone; yet, the secondary osteons and the interstitial bone both have different mechanical properties, and the cement lines around the osteons are also regarded as sites of weakness. Within the osteon itself, the bone comprises sheets of bone (lamellae), which are separated by interlamellar cement lines—both the former and the latter have their own material properties. Even at the most basic level of bone structure (i.e., at the level of the hydroxyapatite crystals within the collagen matrix), bone can be considered as a composite material (see chapter 1).

John Currey (1984, 1991) studied the effects of differences in mineralization on the mechanical properties of bone and found that high values of mineralization correlated with high values of Young's modulus but have low fracture toughness. The opposite is true for bone with low levels of mineralizations. In the living animal, there are various causes for differences in the mineralization of bone. For example, during bone formation, after the deposition of the organic matrix of bone, there is a time delay before the bone mineral is deposited (i.e., the mineralization lag time), which means that there is a difference in the mechanical properties of newly formed bone and "old" bone. Bone remodeling, which involves the removal of primary bone tissues, increases the porosity of the bone and reduces the mineral content of the bone, which also affects its mechanical properties. Thus, variations in mineral content of an animal's bone results from normal biological processes. However, metabolic bone diseases, such as rickets or osteomalacia, also result in depletion of bone mineral (Martin and Burr 1989). Unfortunately, because of diagenesis, the degree of mineralization cannot be assessed in fossil bones. However, by applying engineering principles of beam theory, the strength and stiffness of long bones can be estimated.

At the tissue level of bone structure, it is apparent that different tissue types afford different mechanical properties to bone. The main difference between compact and cancellous bone lies in the porosity of the bone types: the porosity of cancellous bone (i.e., spaces between trabeculae) ranges from 30% to 90%, while the porosity of compact bone (i.e., the osteocyte lacunae, canaliculi, and channels for blood vessels and erosion cavities) is only between 5% and 30%. This therefore means that cancellous bone is substantially less mineralized than compact bone, which affects its material properties (Carter and Spengler 1978). The architectural design of cancellous bone (i.e., the orientation of the trabeculae) and its structural design contribute to its mechanical properties. Compact bone tends to become weaker and more brittle with age as a result of increase in porosity (e.g., seventy-five-year-old people have 250% more resorption spaces than twenty-five year olds; Martin and Burr 1989).

Several researchers on modern vertebrates have demonstrated that the organization of a bone (i.e., in terms of its cross-sectional geometry and distribution of bone tissue) is adapted to the weight of the animal and to the biomechanical properties of the bone itself. Tensile and fatigue tests have shown that the different tissue types in the compacta exhibit different strengths. Thus, the mechanical properties of cortical bone are influenced by its porosity and by its composition. The mechanical properties of woven-fibered bone and circumferential lamellar bone have not been well studied, though it appears that woven-fibered bone without primary osteons is probably the weakest bone type, while circumferential lamellar bone is probably the strongest bone type (Carter and Spengler 1978). In experimental studies on bovine bone under tension, John Currey (1984) found completely remodeled bone to be about 35% weaker than primary bone. Haversian remodeling of primary compact bone decreases bone density and results in reduced tensile strength and fatigue resistance (Carter and Spengler 1978). Since the early 1960s, Antonio Ascenzi and Ermanno Bonucci conducted experimental studies to assess the mechanical properties of secondary osteons. Their various studies have shown that secondary osteons vary in terms of collagen-fiber orientations and are individually designed to overcome particular types of stress that the bone is subjected to.

Bone is recognized as an adaptive tissue that is able to remodel in response to changes in its mechanical environment. When an animal is alive, its bones respond to repeated stresses and strains applied to it by altering its structure, which consequently affects its mechanical properties. For example, in an experimental study examining mammalian limb responses to loading, Daniel Lieberman and Osbjorn Pearson (2001) found that exercised sheep showed higher periosteal modeling and Haversian remodeling rates than sedentary control animals. In the study that Matthias Starck and I had undertaken to examine bone depositional rates in the Japanese quail, we also analyzed the effect of asymmetrical loads (0.7-g weights attached to left or right hip) on the limbs of the birds (Starck and Chinsamy 2001). We found that the femur responded to this loading by altering its normal cylindrical pattern to form asymmetrical cross sections and also increased its bone deposition by two to three times. In a study on pigs, Woo et al. (1981) found that cortical thickness of limb bones of exercised pigs was higher than that of nonexercised ones. Lieberman (1996) obtained similar results on pigs and armadillos. Bertram and Swartz (1991) note that the new bone in the exercised pigs was mainly formed endosteally, when instead it would have been mechanically more optimal for it to form periosteally. They suggest that adaptations in bone may not only be in response to stress exerted, but perhaps additional as yet unknown factors also affect the response of the bone (Bertram and Swartz 1991). These findings complicate the use of fossil bone architecture for interpreting functional adaptations of the bone (Thomason 1995), but perhaps further research on extant animals on the adaptation of cortical bone to its mechanical environment will shed more light onto these problems.

The actual cross-sectional shape of long bones has been extensively used by physical anthropologists in calculating the forces that acted on the bone and permit calculations of the absolute and relative bone strength. In a study of a growth series of femora of *Dryosaurus lettowvorbecki*, a herbivorous dinosaur from Tanzania, Ron Heinrich and his colleagues Christopher Ruff and David Weishampel of Johns Hopkins University (1993) found a distinctive change in the relative amount and distribution of cortical bone through ontogeny. They ascribed these changes in the cross-sectional properties of the femora to increasing mechanical loads and a caudal shift in the animals' center of gravity as a result of a change in its preferred mode of locomotion. From these studies, they deduced that juvenile *Dryosaurus* were quadrupedal and that later in ontogeny they shifted to bipedality.

From the above, it is apparent that a biomechanical assessment of the structure of fossil bones can provide some clues about the mechanical properties and function of the particular skeletal element when the animal was alive. By studying the overall geometry of the bones and the cross-sectional geometry of the bones and by comparisons with modern taxa, the range of forces that act on it can be assessed (Thomason 1995). The problem, however, is that such analyses require data such as weight estimates, soft tissue (muscle reconstructions), bone strength, and force estimates—all of which are not readily available from the fossil animal. Analysis of the distribution of bone microstructure (i.e., cancellous vs. compact, distribution of secondary osteons, location of different tissue types) can also provide valuable information about the possible forces that the bone had to overcome during life.

Bone Wall Thickness

A functional relationship is recognized between bone wall thickness, the habitat of the animal, and the forces that act on the bone. Experimental studies by researchers John Currey and McNeil Alexander (1985) have suggested that vertebrates' long bones have minimum mass, which enables them to perform locomotor functions at minimum energy cost.

Cross-sectional architecture of femora of different species reveals distinct differences in bone wall thickness, which appear to be related to the lifestyles of the animal. Some bones have thick bone walls and a small medullary cavity or sometimes none, while others have thin bone walls and large medullary cavities. Much research has been directed to quantify and ascertain why these differences occur. In a study examining the structure of bird bones, German researcher Paul Bühler (1986) quantified observable differences in bone wall thickness for a variety of birds. He deduced that long bones of birds are highly developed lightweight structures, with large flying birds having relatively much thinner bone walls than medium-sized and small bird species.

In an attempt to determine the use of bones for ascertaining the life habits of extinct mammals, William Wall (1983) measured the volume and weight of limb bones of a variety of extant terrestrial and aquatic mammals. He found that the limb bones of aquatic animals exhibited higher densities than terrestrial ones and suggested that aquatic animals use an increase in compact bone mass to overcome buoyancy in water.

In the study of the relative bone wall thickness (RBT) of *Syntarsus, Massospondylus, Crocodylus niloticus,* and two species of birds, I found that the amphibious crocodile exhibited the highest bone wall thickness, while both the secretary bird (*Sagittarius serpentarius*) and the ostrich (*Struthio camelus*) had much lower RBT than the dinosaurs. The secretary bird, a large bird capable of flight, had the lowest RBT, which contributes to its weight reduction. However, penguins have extremely thick bone walls that helps them overcome buoyancy in the water (Fig. 3.11; see Chinsamy et al. 1998). In several bird species, bony trabeculae (particularly in the proximal and distal regions) run across the hollow medullary cavity, and it has long been suggested that these act as "beams" and afford strength to the bone. In an experimental study on the mechanical properties of the humerus of pigeons (*Columba livia*), Rogers and LaBarbera (1993) found that the trabeculae significantly improved the flexibility, rigidity, strength, and toughness of the bone.

In their analysis, Currey and Alexander (1985) found that *Pteranodon* had remarkably thin bones even when compared with a range of flying birds. They further deduced that in *Pteranodon* reduction in the mass of the bone appears to be a very important factor, which means that the bones are not as sturdy as bird bones that would be able to withstand localized impact. In a recent study of pterosaur bone microstructure, Armand de Ricqlès and colleagues (2000) report that pterosaurs have the thinnest bone walls among tetrapods. Long bones with lengths of about 100 mm typically have bone walls that are about 0.5 mm or less, while smaller bones can have bone walls that are much thinner. The researchers note also that for comparably sized individuals pterosaur bones are thinner than that of birds. As in many avian taxa, pterosaurs also have bony struts that traverse the medullary cavity, and it is likely that these also contribute to the mechanical strength of the bone.

I have found significant variation in the cross-sectional geometry of the bones of various Cretaceous birds, which appear to be dependent on the lifestyle that the bird is adapted to. For example, I found that the relative bone wall thickness of the volant enantiornithines is about 13.7%, while that of the flightless, terrestrial *Patagopteryx* is 18%, and aquatic birds had much thicker bone walls (Chinsamy et al. 1998).

In a study examining allometry and adaptations in the long bones of a digging group of rodents (Ctenomyinae), Casinos and colleagues (1993) found that digging induced changes in the morphology of the long bones. They found that specialized digging rodents have shorter and relatively thicker long bones. Recent work on the bone microstructure and relative bone wall thicknesses of several nonmammalian therapsids suggest that *Diictodon* (Ray and Chinsamy 2004) and *Trirachodon* (Botha and Chinsamy 2004) show adaptations for digging, which suggests that they may have had a fossorial lifestyle.

Bone Microstructure of Aquatic Vertebrates

Aquatic vertebrates often show gross morphological adaptations for their particular habitats, but because soft tissue is rarely preserved in the fossil record, our under-

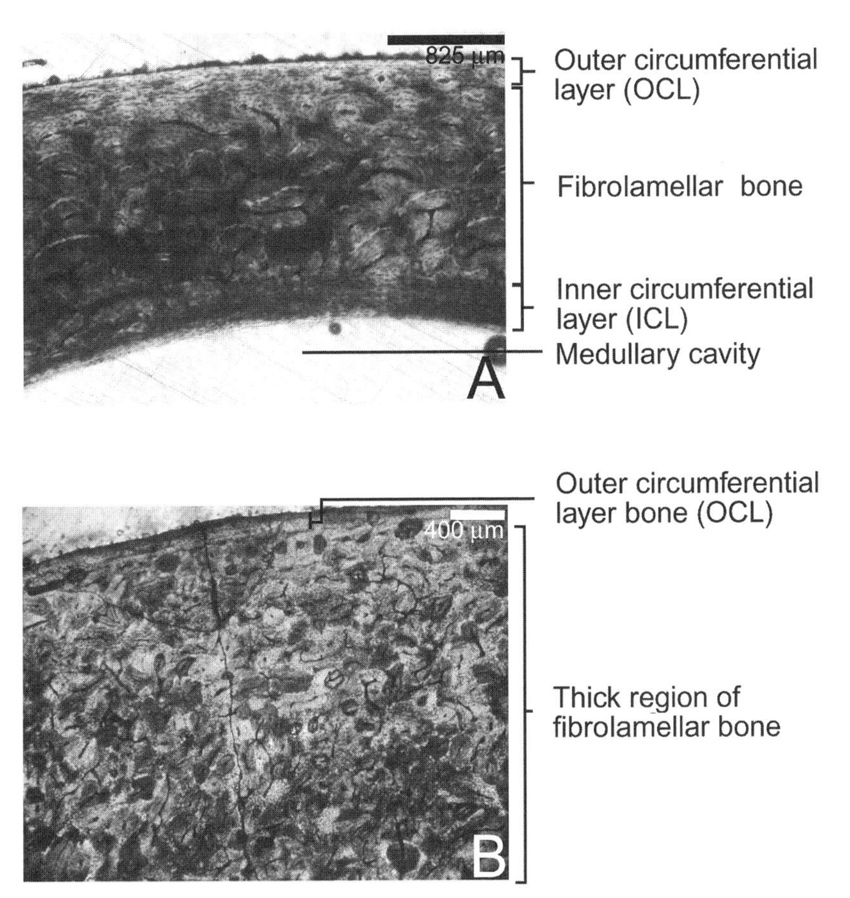

Fig. 3.11. Comparison of the bone wall thickness in transverse sections femora of a modern flying bird and a flightless diving bird. *A*. An albatross, showing the typical three-layered structure of many modern birds with the outer circumferential layer (OCL), middle fibrolamellar bone, and endosteal or inner circumferential layer (ICL). *B*. A penguin. Note the relatively thin wall of the flying bird and the uninterrupted nature of both the bird bones (i.e., absence of lines of arrested growth).

standing of aquatic fossil vertebrates mostly stems from an analysis of their bones. Besides modifications in bone morphology, research on bone microstructure of both modern and fossil aquatic vertebrates has shown that the internal organization of the bone is often similarly modified (in a convergent or parallel manner, irrespective of phylogenetic affinities) to accommodate the aquatic lifestyle of the animal. French researcher Vivian Buffrénil and his colleagues have shown that depending on the particular type of adaptation of the animal the bones can be transformed into either a very compacted, dense structure, through pachyostosis (osteosclerosis or pachyosteosclerosis) or into an extreme cancellous (spongy, highly porous) bone by osteoporosis (e.g., Buffrénil and Shoevaert 1988; Buffrénil 1989; de Ricqlès and Buffrénil 2001). The resulting bone types are referred to as *pachyostotic* bone and *osteoporotic* bone.

The pachyostotic type of adaptation is evident in nothosaurs and sirenians, while the highly porous, cancellous osteoporotic bone has been observed in ichthyosaurs, as well as in modern cetaceans and marine turtles. Buffrénil and his colleague Mazin (1989) used the pachyostotic nature of the compacta of *Claudiosaurus germaini* to suggest that it was an aquatic reptile, while Amy Sheldon (1977) used the differences in density of mosasaur rib microstructure to suggest that *Platecarpus* lived in shallower waters and *Clidastes* and *Tylosaurus* inhabited deeper waters.

Definite anatomical adaptations, in response to their aquatic lifestyles, are seen in the Cretaceous birds *Hesperornis* and *Polarornis,* the Antarctic loon. My studies of their bone microstructure show that they tend to have pachyostotic bones (i.e., they have relatively thick bone walls that suggest adaptations for diving (Fig. 3.11). For example, the average bone wall thickness of *Polarornis* is 37% that of its diameter, which is much thicker than that of the modern loon, *Gavia stellata,* in which the relative bone wall thickness is only 15% (Chinsamy et al. 1998). However, the thickness of *Polarornis* bone appears to be more comparable with that of modern penguins, such as the emperor penguin, which has a femoral bone wall thickness of 33% and the Humboldt penguin, which measured 31% (Fig. 3.11). It is possible that the difference in relative bone wall thickness of aquatic birds reflects their varying ability to dive to different depths.

Conclusions

The microscopic structure of bones provides a host of information about how they grew and about factors that affected their growth during ontogeny. There appears to be no unambiguous bone depositional rate that results in a particular type of bone tissue being formed. Preliminary studies suggest that much variation exists and that many factors affect the rate at which bone forms—among them are ontogenetic age, location in the skeleton, location in a particular skeletal element, as well as a broad range of environmental factors.

The nature and distribution of zonal bone versus azonal (uninterrupted) bone among vertebrates shows that although reptiles typically form zonal bone and mammals form azonal bone, this does not always hold true. Many exceptions occur, and it is increasingly evident that bone shows substantial plasticity and is able to respond to a variety of environmental factors.

New developments in the field have suggested that the shape of the channels in bone have no bearing on the shape of the actual blood vessels. Although the existing terminology can still be used, researchers should understand that they refer to the shape of the channels in the bone (and not to the shape of the blood vessels). Studies of the bones of modern vertebrates show the enormous scope for using the microscopic structure of fossil bones to reconstruct various aspects of the biology of an extinct animal.

CHAPTER FOUR

Inside Dinosaur Bones

Pioneering studies of dinosaur bone microstructure can be traced to about the mid-1800s. By this time, largely because of the work by seventeenth-century researchers on bone Clopton Havers and Anton van Leeuwenhoek, and the efforts of a number of eighteenth-century researchers, the scientific community was well aware of the tissue-level organization of modern bone and also recognized that Haversian systems are formed by the resorption of primary bone and result from the secondary redeposition of bone.

As early as 1849, John Thomas Quekett proposed the use of bone in the identification of fossil taxa and noted that the bones of *Iguanodon* did not differ substantially from those of modern lizards. Quekett was the first to recognize osteocyte lacunae in the bones of *Iguanodon,* which he referred to as *bone cells*. Soon after, Gideon Mantell (1850) identified Haversian canals and "bone cells" in the sauropod, *Pelorosaurus,* and illustrated osteons in the dorsal spine of *Hylaeosaurus.* In 1859, Sir Richard Owen (who is famous for having named the Dinosauria) proposed that sauropod bone structures were evidence of their aquatic mode of life and also figured a vertebrae of *Cetiosaurus* that shows dense laminar or plexiform tissue. In 1878, Hasse used *Thecodontosaurus* skeletal material to illustrate endochondral ossification in fossil reptiles, and he contrasted the occurrence of spindle-shaped bone cells in the periosteal bone with stellate examples in a nothosaur.

In these early studies, researchers were simply undertaking cursory examinations of dinosaur bones, without recording many microstructural details. The first real systematic undertaking of dinosaur bone microstructure was by Adolf Seitz (1907). He illustrated the histology of several dinosaurs (*Megalosaurus, Iguanodon, Trachodon* [*Anatosaurus*]), and he also made reference to the actual bone cells, "knochen-köperchen," in several dinosaurs (e.g., in *Plateosaurus, Megalosaurus, Brontosaurus* [*Apatosaurus*], *Diplodocus, Camarasarus, Haplocanthosaurus, Allosaurus, Stegosaurus, Iguanodon, Dryptosaurus, Triceratops,* and *Trachodon* [*Anatosaurus*]). In this publication, Seitz also described what he thought was fossil blood corpuscles, "fossilen Blutkorperchen," in the bones of *Iguanodon bernissartensis,* which were later rejected by Swinton (1934). Seitz is known for having firmly established the distinction of primary and secondary vascular canals in compact bone. He also provided the first illustrations of zonal bone in dinosaurs from *Allosaurus* and *Stegosaurus.*

In 1916, J. S. Foote was the first to demonstrate that many modern animals have compact bone structure that was different from human bone, and he developed a nomenclature of fossil bone based on gross structure. Foote recognized three structural units in bone: lamellae (lamellar), laminae (laminar), and Haversian systems. However, these terms have not been used consistently by various authors and have resulted in tremendous confusion. For example, Foote's lamellar bone refers to gross structure, and is not the same as the modern usage of the term, which refers to the fibrillar organization of the bone. Instead, Foote's lamellar bone is equivalent to Reid's "lamellated" bone (Reid 1996). Another example of conflicting use of terminology is Foote's "laminae" or "laminar" bone, which he used to refer to concentric vascular networks, but today this could refer to the vascular zones of crocodilians or the laminar fibrolamellar bone of modern mammals (Reid 1996). Foote also recognized a complicated suite of different types of Haversian bone on the basis of whether they were fully differentiated, and he included both primary and secondary osteons in his descriptions. In 1922, Broili described the microstructure of ossified tendons of a hadrosaur, and about two years later, Moodie described ossified tendons in *Anatosaurus* (now known as *Edmontosaurus*) and *Ankylosaurus.* Moodie (1924) also reported similar structures to Seitz's (1907) "fossilen Blutkorperchen" in a foot bone of *Apatosaurus* in close proximity to Howship's lacunae, which Swinton (1934) saw as evidence of iron-stained osteoclasts. Moodie (1923) is particularly well known for having recognized a range of pathological conditions in dinosaurs.

In 1933, the landmark ontogenetic study by Baron Francis Nopsca and his colleague Heidsieck was published in which they used bone microstructure to distinguish between the ontogenetic status of half-grown and immature hadrosaurs. By this time, substantial progress had been made regarding understanding the fibrillar organization of modern bone, and Walter Gross (1934) took advantage of the progress made by

Weidenreich (1930) with regard to the fibrillar organization of bone. Gross used bones of *Plateosaurus* and *Brachiosaurus* to develop concepts of laminar bone and the functional significance of primary and secondary osteons. At about the same time, other reports of dinosaur bone microstructure were made by Peterson (1930), von Eggling (1938), de Lapparent (1947), and, later, Cook and colleagues (1962).

Between 1956 and 1958, Americans Donald Brown and Sidney Enlow published a series of papers that documented and provided a better understanding of histological patterns among major vertebrate groups, from fishes to mammals, including both fossil and recent samples. In this study, they examined the microstructure of several dinosaurs, noting that a distinctive feature of dinosaur bone was the "relative intensity of secondary tissue replacement" (Enlow and Brown 1957, p. 200). They provided illustrations of the bone microstructure of *Brachiosaurus, Plateosaurus, Stegosaurus* (which they obtained from Gross), as well as that of *Allosaurus, Megalosaurus, Dryptosaurus, Diplodocus,* and *Iguanodon* (which they obtained from Seitz). Frank Peabody working in California had begun studying growth marks in bones but his untimely death in 1958 halted further research. Fortunately, his daughter and colleague managed to publish his findings a few years later (Peabody 1961).

John Currey (1960), who is extremely well known for his work on the mechanical properties of modern bones, also contributed to understanding dinosaur bone tissues and the vascularization of different bone tissue types. In 1962, he described the bone tissue of two prosauropod dinosaurs from the Forest Sandstone of Zimbabwe and noted that the bone was richly vascularized—more so than that of recent reptiles and equivalent or more than that of recent mammals of comparable sizes. Currey postulated that perhaps the high levels of vascularization observed in the dinosaur bones are related to physiological specializations in the dinosaurs that are greater than that of recent reptiles (Currey 1962).

By about the mid-1960s, Roman Pawlicki began his search for biomolecules (e.g., collagen, nucleic acids, and lipids) in the bones of *Tarbosaurus* from the Gobi Desert. Over the next several years, alone and with various colleagues (e.g., 1966, 1977, 1983, 1987), he published a host of papers in this area and on scanning electron microscopy analysis of dinosaur bone tissues.

In 1965, John Campbell published a fascinating account of a lesion in a dinosaur bone from the Upper Cretaceous of Transylvania that showed a pathological bone microstructure similar to that seen in avian osteopetrosis. In his detailed review of the bone microstructure of reptiles, Donald Enlow (1969) described the bone microstructure of various recent and fossil reptiles. He illustrated a *Triceratops* rib showing secondary reconstruction and also notes that extensive Haversian remodeling is a general feature of both ornithiscians and saurischian dinosaurs. Enlow specifically mentioned that plexiform bone tissue occurred as primary bone tissue in *Brachiosaurus* and *Plateosaurus* but becomes replaced by secondary reconstructions.

In the late 1960s and early 1970s, the French comparative anatomist, Armand de Ricqlès, advanced the field of paleohistology substantially by documenting bone structure among a variety of extinct tetrapods, including a wide range of nonmammalian therapsids (e.g., de Ricqlès 1972, 1975) and dinosaurs. De Ricqlès (e.g., 1974, 1976, 1980) reported in a host of publications that dinosaur compact bone tissues largely consisted of fibrolamellar bone and Haversian bone and were therefore unlike that of "typical" lamellar-zonal bone common in reptiles and more like bone tissues observed in fast-growing mammals and birds (this assertion was later contested in 1981 by Robin Reid). Armand de Ricqlès is still actively involved in research and his colleagues in Paris, Jacques Castanet, Vivienne de Buffrénil, and Helene Francillon, have all done (and are still doing) exciting work in the field of bone microstructure of both extinct and extant vertebrates.

In the 1980s, Robin Reid, an invertebrate paleontologist by training working in Belfast, switched to working on dinosaur bone microstructure and, in so doing, provided a fresh perspective on the interpretation of dinosaur bone microstructure. Reid (1981) is often cited as the first person to recognize zonal bone tissue in an undescribed sauropod dinosaur. However, it appeared that, in fact, Campbell (1966), in the publication describing a dinosaur bone lesion, also described (and illustrated) "several cortical annular lamellae" in the dinosaur bone that he recognized as a general feature of both fossil and modern reptilian long bones and suggested further that these appear to indicate "growth periodicity." It therefore appears that almost two decades before Reid's announcement of zonal bone in a sauropod pelvis and about the same time that de Ricqlès was undertaking studies of dinosaur bone tissues, Campbell (1966, p. 167) had recognized "zonal" bone in the cortical bone tissue of a dinosaur. I suspect that because this paper appears to deal with pathological bone, the description of cortical zonal bone (described by Campbell as "annulate structure of cortical bone" and "cortical annular laminae") remained inconspicuous. Campbell (1966) also described rest lines as being "close together" near the periosteal surface and says that they "probably correspond to the outer circumferential lamellae of mammalian bone."

In 1981 when Reid announced his find of zonal bone in a sauropod dinosaur, this publication commanded attention because it came at a time when dinosaur bone was believed to be unlike the bones of reptiles and was formed as a result of rapid osteogenesis as seen in endothermic mammals. Subsequently, Robin Reid published several articles relating to dinosaur bone microstructure and dinosaurian physiology (e.g., Reid 1984a, 1984b, 1987, 1990, 1993, 1997a, 1997c) and dinosaur growth (e.g., Reid

1997b). An essential reading and ideal reference for students of fossil bone tissues is Robin Reid's (1996) "Bone Histology of the Cleveland-Lloyd Dinosaurs and of Dinosaurs in General." Besides the detailed descriptions of various tissue types that can be encountered in dinosaur bone, one of the remarkable aspects of this synthesis is the clarification of conflicting terminology used by various researchers in the past and their current usage. For example, in Reid (1996) and in Gross (1934), laminar bone is equivalent to Enlow's (1969) plexiform bone and encompasses de Ricqlès' (1974) plexiform and laminar bone types.

In 1987, Peter Houde published the first account of bone tissues of the Mesozoic diving bird, *Hesperornis.* My own work, since the late 1980s, deals with the histological changes through ontogeny among dinosaurs: *Syntarsus rhodesiensis* (Chinsamy 1990); *Massospondylus carinatus* (Chinsamy 1993a, 1993b); *Dryosaurus lettowvorbecki* (Chinsamy 1995b); and other extinct and extant archosaurs, for example, *Postosuchus* (Chinsamy 1994); *Crocodylus niloticus* (Chinsamy 1991, 1993b); as well as with the bone microstructure and biology of Early (Mesozoic) birds, for example, *Rahonavis,* several enantiornithines, *Patagopteryx, Vorona,* and *Hesperornis* (e.g., Chinsamy et al. 1994; Chinsamy and Elzanowski 2001; Chinsamy 2002). The nature of my work is largely comparative, and I am interested in deciphering biological signals recorded in the bones of living birds and reptiles, which could then be applied to fossil animals. At my laboratory in South Africa, my students and I have worked on a variety of modern animals, dinosaurs, and fossil animals other than dinosaurs. For example, people in my lab have worked on extant sharks, crocodiles, lepidosaurs, a variety of birds, ungulates, microvertebrates from Langebaanweg, extinct and extant girrafids, a variety of nonmammalian synapsids, and pterosaur bones and teeth. Because of various constraints, paleontology is not a popular career option in South Africa. However, there is a small but steady stream of paleontology graduates from South Africa, particularly from the University of Cape Town and from the Bernard Price Institute for Palaeontological Research in Johannesburg. Doctoral graduates from my lab are Tamara Franz-Odendaal, now at Brian Hall's Evolutionary Development laboratory in Canada, and Jennifer Botha, now at Iziko Museums of Cape Town. Sanghamitra Ray, now in Calcutta, spent two years at my laboratory as a postdoctoral fellow working on nonmammalian synapsids. I co-supervised Argentinian Leonardo Salgardo's PhD work on sauropod bone microstructure and have also worked closely with Allison Tumarkin-Deratzian of the University of Pennsylvania (now at Vassar College), who worked on bone texture and microstructure of extant birds, crocodilians, and dinosaurs.

Other contemporary workers on dinosaur bone microstructure are Jack Horner of the Museum of the Rockies in Bozeman, Montana, and Kevin Padian of the University of California, both of whom have been involved collaboratively with Armand de Ricqlès for several years. This group of researchers has recently published several articles dealing with bone microstructure of dinosaurs: for example, *Hypacrosaurus* (Horner et al. 1999), *Maiasaura peeblesorum* (Horner et al. 2000a), *Confuciusornis* (de Ricqlès et al. 2003). Former students from this research group are Mary Rigby-Schweitzer who worked on *Tyrannosaurus rex* bone and reported on their possible red blood cells (later suggested to be pyrite by Martill and Unwin 1977); David Varricchio (1993), who worked on multiple skeletal elements of a growth series of *Troodon formosus;* Kristi Currey-Rogers (1999), who studied bone microstructure and growth of *Apatosaurus;* and Greg Erickson, who with Russian scientist Tatyana Tumanova worked on a growth series of femora of *Psittacosaurus* and in 2001, with Kristi Currey-Rogers and other colleagues, worked on dinosaurian growth rates. More recently, Greg Erickson and co-workers (2004) have published on the growth dynamics of tyrannosaurids. Over the past few years, Rigby-Schweitzer (e.g., 1997) has invigorated the earlier work of Roman Pawlicki in search of biomolecules (e.g., nucleic acids, proteins, collagen, lipids, etc.) in dinosaur bones using sophisticated immunochemistry techniques.

Besides the labs mentioned above with focused research on bone microstructure, there have been several independent workers on bone microstructure of dinosaurs, for example, Madsen (1976), who illustrated a bone section of *Allosaurus fragilis* and the work on the Proctor Lake ornithopods by Dale Winkler (1994). In 1995 Frédérique Rimblot-Baly, working with Armand de Ricqlès and Louis Zylberberg, published an analysis of the bone microstructure of a growth series of the Madagascan sauropod, *Lapparentosaurus madagascariensis.* In this paper, Rimblot-Baly used the cortical thickness of the bone, divided by the number of growth rings present in the bone, to work out the yearly growth rate, and from this, they calculated the daily bone depositional rate. John Rensberger and Mahito Watabe (2000) published a paper on the fine structure of bone in dinosaurs and mammals. Herein they suggest that the canaliculi and collagen-fiber bundles were differently organized in different dinosaur lineages, with the ornithomimid dinosaurs having an arrangement similar to birds, while ornithiscian dinosaurs were more like mammals with regard to the features mentioned (Rensberger and Watabe 2000). I am not convinced by this because, in my experience, the orientation of the canaliculi are highly variable and tend to be age- and rate-related, and they also can appear quite different, depending on how the bone is sectioned. Perhaps more research on the fine structure of the canaliculi on both extant vertebrates and dinosaur taxa might clarify whether the reported differences are real. Also, in 2000, Martin Sander, based in Germany, published his findings on the bone microstructure of the Tendaguru sauropods (Sander 2000). Sander describes "polish lines," which are growth lines visible with the naked eye in polished sections of bone (not thin sections) and appear to be the same lines that Armand de Ricqlès (1983) used to estimate the age of *Bothriospondylus.* More recently, Sander with Christian Tückmantel (2003) published a paper dealing with age estimation of sauropod limb bones based on lamina thickness in the cortical fibrolamellar bone.

A Synthesis of Nonavian Dinosaur Bone Tissues

Upon sectioning a dinosaur bone, compact (cortical) and cancellous (spongy) bone are readily discernible. Most dinosaur long bones appear to have a fairly thick bone wall made up of compact bone tissues.

Under a light microscope, several types of compact bone tissues (e.g., primary periosteal bone, accretionary bone, Haversian bone, and metaplastic bone) can be discerned. Cancellous bone usually occurs internally, around the medullary cavity, and at the proximal and distal ends of long bones. In flat bones, the cancellous internal bone is also referred to as *diploë* (Reid 1996). Several types of cancellous bone are also recognizable in dinosaurs (e.g., endochondral primary bone, secondary reconstructed cancellous bone, dental bone, and cavernous bone). Without a consideration of the actual nature of the bone tissues, relative proportion of the amount of cancellous and compact bone can be readily deduced, and relative bone wall thickness can be calculated. Cross-sectional geometry of the bone can also be assessed at this stage (Heinrich et al. 1993).

Compact Bone Tissues

Perisoteal (Primary) Bone

Microscopic study of thin sections of dinosaur bones generally reveal excellent preservation of bone microstructure. As mentioned earlier, osteocyte lacunae (referred to as *bone cells*) in dinosaur bones have been recognized since 1849 and shortly thereafter, different bone microstructure (tissue) types were recognized on the basis of fibrillar organization.

Considering that the channels observed in the bones of dinosaurs housed blood vessels and various connective tissues, the number of channels have often been used as a proxy of the vascularization of the compacta (though this is not accurate since Starck and Chinsamy (2002) have shown that in juvenile modern bird bone only 20% of the channel is occupied by blood vessels). Nevertheless, abundant channels in dinosaur compact bone tissues are indicative of relatively high bone vascularization. Most compact bone in dinosaurs appears to be of the fibrolamellar type in which primary osteons are embedded in a woven bone matrix (Fig. 4.1). Under polarized light, birefringence patterns (axial cross) of primary and secondary osteons are readily distinguishable (Fig. 4.2).

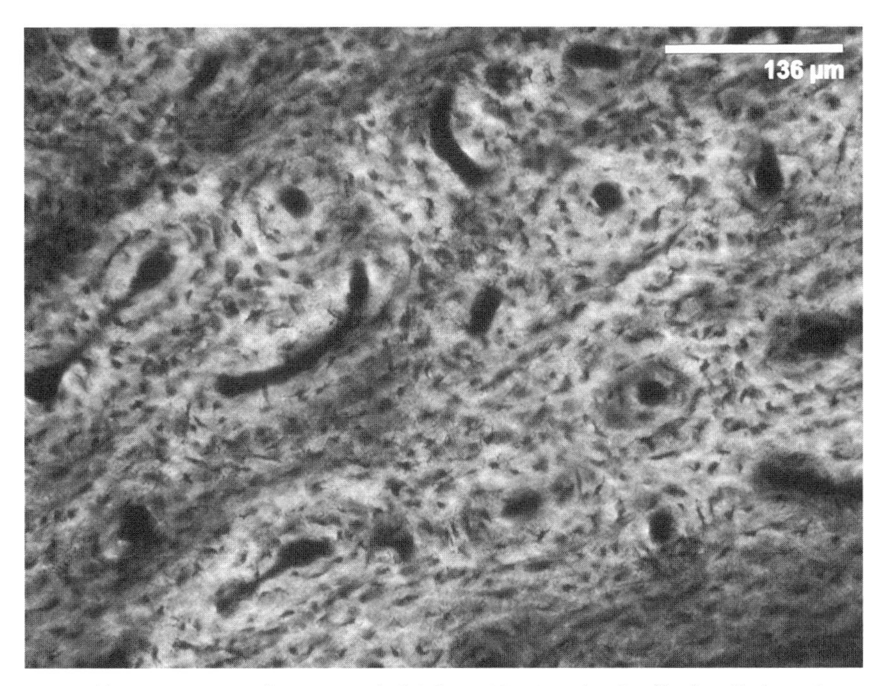

Fig. 4.1. Transverse section of a metacarpal of *Archaeornithomimus* showing fibrolamellar bone tissue.

For a long time, and especially because of the work of de Ricqlès (e.g., 1974, 1976), dinosaurs were considered to have extensively vascularized fibrolamellar bone tissues that are often replaced by extensive development of Haversian bone through secondary reconstruction. This was sharply contrasted with the more poorly vascularized zonal bone tissues of reptiles (e.g., de Ricqlès 1974; Fig. 3.5A, 3.5B). It is true that dinosaur bone tissues are highly vascularized, and the extensive development of fibrolamellar bone and Haversian bone is quite distinctive, but there is much more to dinosaur bone than this.

As mentioned earlier in this chapter, a description of periodically formed "annulate structure" of cortical bone in a dinosaur (similar to extant reptiles) was described as early as 1966 by Campbell, but for some reason, this description did not make it into the mainstream literature on dinosaur bone tissues. However, Robin Reid's (1981) announcement in *Nature* of the occurrence of lamellar-zonal bone in the pelvis of a sauropod dinosaur brought to the fore the fact that not all dinosaurs form bone in an uninterrupted sustained manner. It is curious that in the article, Reid notes that the most similar bone tissue to that which he describes in the sauropod is from an Early Triassic amphibian, *Stenotosaurus*. De Ricqlès (1983) subsequently reported cyclically formed bone in the humerus of *Bothriospondylus,* and a few years later, Robin Reid (1990) documented zonal

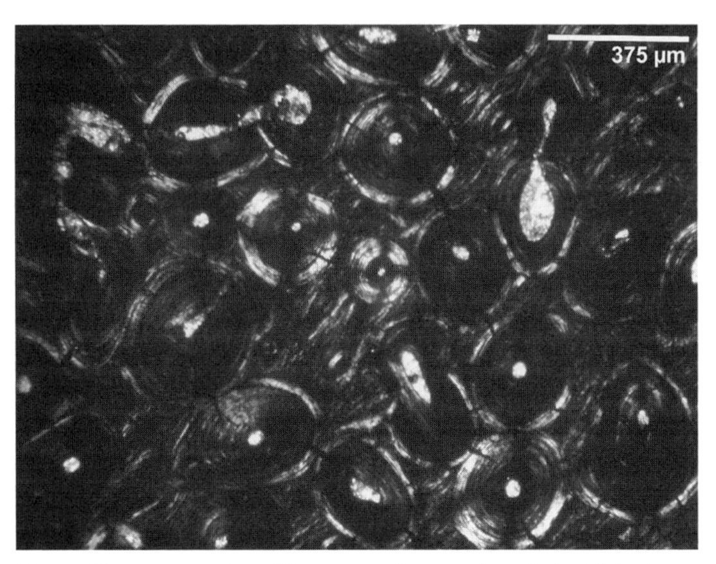

Fig. 4.2. Transverse section of *Kentrosaurus* femur under polarized light showing the fine light and dark banding typical of lamellar bone. Note also the "axial crosses" that are distinctive of the Haversian systems.

bone tissues in a wide variety of dinosaurs representing prosauropods, sauropods, theropods, ornithopods, a stegosaur, and a ceratopsian. A series of other studies have documented zonal bone among several dinosaur taxa (e.g., Chinsamy 1990, 1993a; Varricchio 1993; Fig. 4.3; Plate 5A). Thus, contrary to previous understanding, zonal bone is apparently fairly widely distributed among various dinosaur genera, although some genera are capable of forming uninterrupted fibrolamellar bone tissue.

Zonal bone among many dinosaur taxa usually takes the form of alternating fibrolamellar bone and lines of arrested development (LAGs), with or without annuli (Fig. 4.3; Plate 5A). As in extant vertebrates, the LAGs indicate a pause in osteogenesis, while the annuli consist of a more slowly formed bone, such as lamellar or lamellated bone. Reid (1990) described several dinosaurs that show closely spaced rest lines within the annulus (e.g., in a *Baryonyx* rib, and in an *Allosaurus* ungal and radius). Thus, in several dinosaur taxa, fibrolamellar bone is periodically formed and, possibly, annually (i.e., bone deposition switches periodically from a rapid rate of bone deposition to a slower rate of bone formation). The trigger for this alternating rate of bone deposition is unknown but could be seasonal as in modern reptiles. As in modern crocodilians (which is considered an extant outgroup of the phylogenetic bracket of dinosaurs), as well as in other reptiles, the zones appear to be wider during early ontogeny and then narrow with increasing age.

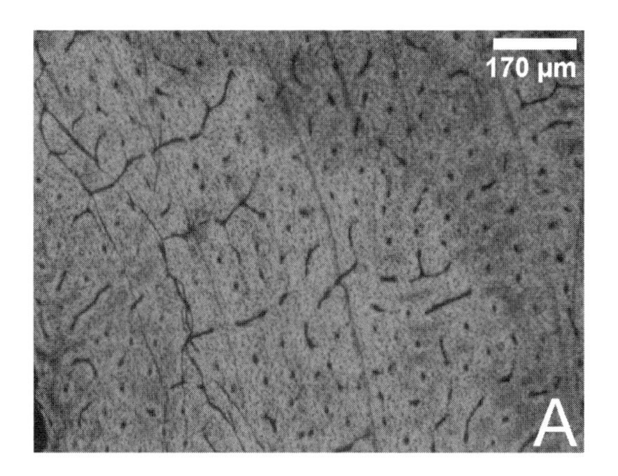

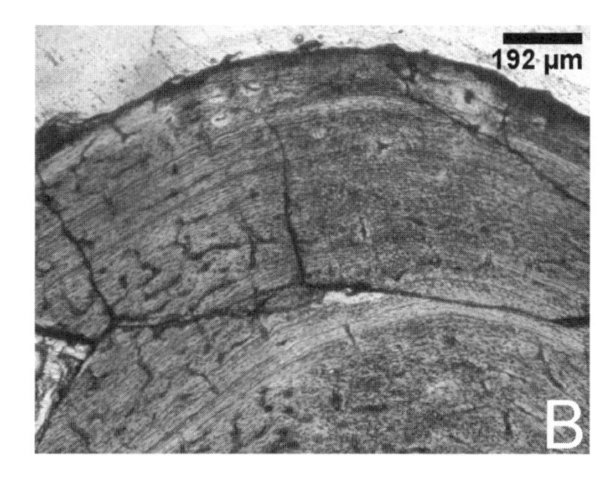

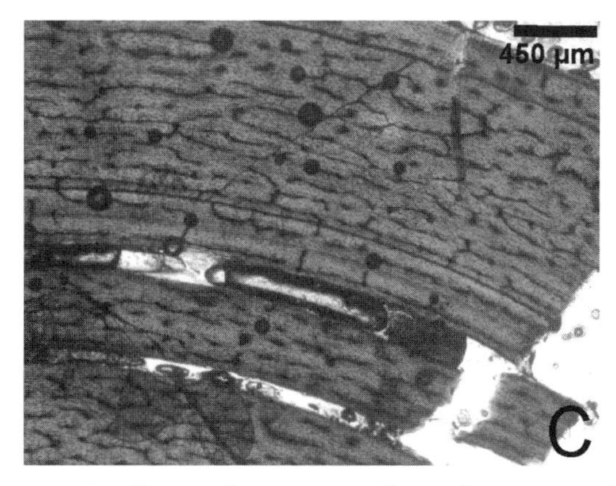

Fig. 4.3. Zonal bone in dinosaurs. *A. Archaeornithomimus. B.* A theropod femur. *C. Unenlagia* humerus.

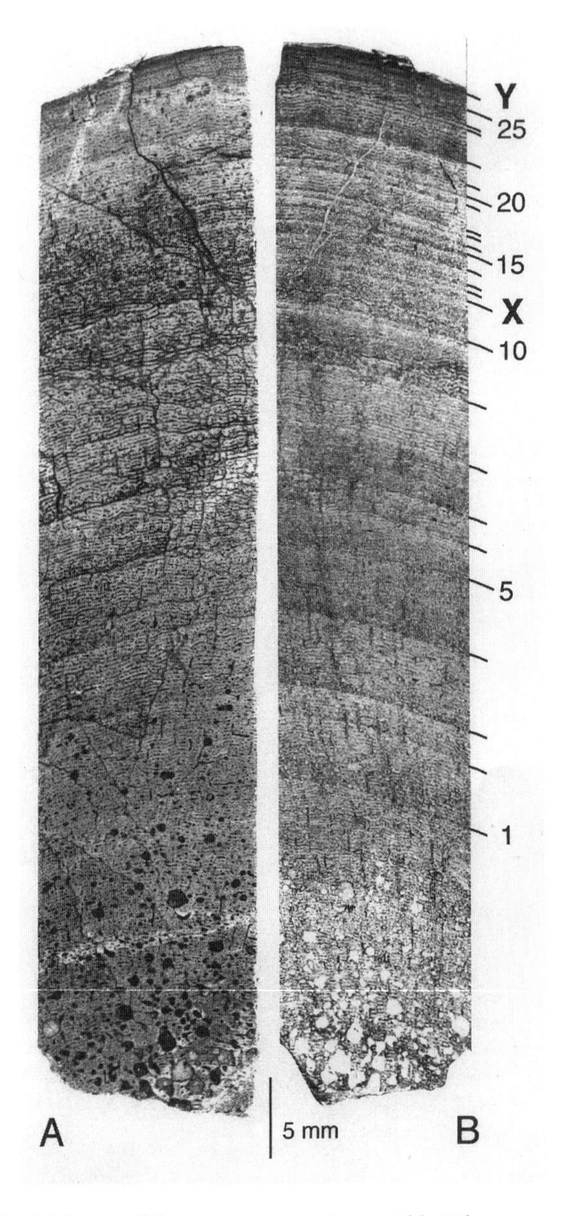

Fig. 4.4. Transverse section of a polished section of *Janenschia* femur core showing the thick laminar cortical bone tissues. *A*. Under normal light. *B*. Under reflected light. Polish lines are indicated by short lines to the side. Note the wide spacing of the first eleven lines. This has been taken to mean that sexual maturity occurred thereafter (after *X*). *Y* indicates the deposition of lamellar-zonal bone, which suggests a slowing down in the rate of bone formation. Image courtesy of Martin Sander.

Among extant reptiles, once sexual maturity is reached, there is a definite decrease in the width of subsequent zones. Martin Sander (2000) found that the first ten growth rings in the Tendaguru sauropod, *Janenschia*, were fairly widely spaced but thereafter, they became substantially narrow (Fig. 4.4). Sander attributed the change in the width of the zones to the attainment of sexual maturity. In my study of *Massospondylus*, I found that the thickness of the zones were quite variable (Chinsamy 1993a), and I was unable to locate any distinctive narrowing of the zones. Because of the variability that I found, I propose that, besides the attainment of sexual maturity, there are a number of other factors (such as nutrition and local growth conditions) that also affect the amount of bone that is deposited at a particular time. In fact, in *Janenschia*, the width of the growth rings also tend to be quite variable (Fig. 4.4), where, for example, growth rings 2–3, 6–7, 21–22, 22–23 are about the same size. Perhaps if large sample sizes are available and there is a consistent reduction in zonal width among similar age classes, at approximately similar "ages," a reasonable case can be made for attributing the change in bone depositional rate to the attainment of sexual maturity.

Identification of zonal bone in isolated dinosaurs should be treated with caution because noncyclical "growth rings" are known from several dinosaur taxa. These "rings" may be caused by other factors such as local changes in growth for morphological reasons (e.g., bone drift). However, if growth series of dinosaur bones show increasing number of growth rings with increasing size, for example, *Syntarsus* (Chinsamy 1990); *Massospondylus* (Chinsamy 1993a); *Troodon* (Varricchio 1993); *Psittacosaurus* (Erickson and Tumanova 2000), it is reasonable to assume that they represent periodic interruptions (which were most likely annual, as in modern reptiles), and skeletochronology can be applied.

However, a potential problem that one needs to be aware of when undertaking skeletochronological assessments is that in a single skeleton of an individual dinosaur different bones may reflect different numbers of "growth rings" (Horner et al. 1999). Such variation in numbers of growth rings in the skeleton is expected because different bones in the skeleton have different morphology, and each requires specific remodeling processes, for example, a neural spine grows differently to a tibia. This was aptly illustrated by Horner and colleagues (1999), where they found that in a single skeleton of *Hypacrosaurus stebingeri* the long bones, the tibia and the femur, had eight growth rings, and the radius and fibula both had seven growth rings, while a neural spine and a scapula preserved only three and five growth rings, respectively. It is apparent that given the variability of growth rings in the skeleton caution is advised in selecting study material. It does, however, appear that long bones, particularly sectioned in the midshaft region, best preserve the growth history of an individual. To assess longevity and growth data for a particular taxon, it would be best to assess a growth series of femora or tibia, although the examination of other bones in the skeleton will provide an indication of histological variability in the skeleton.

Contrary to findings in other dinosaur taxa (e.g., *Hypacrosaurus stebingeri*; Horner et al. 1999), Erickson and colleagues (2004) suggest that cortical remodeling is negligible in nonweight-bearing bones (pubes, fibulae, ribs, gastralia, and postorbitals) and suggest that these skeletal elements are best for skeletochronology in tyrannosaurids. I am skeptical about the premise that weight bearing dictates the amount of remodeling that occurs. Surely, several other factors such as the shape of the bone and how this is accommodated during growth, the extent of secondary reconstruction, lifestyle adaptations, and age are also important considerations. My doubt is enforced by personal observations in a *Tyrannosaurus* fibula (one of the skeletal elements Erickson and colleagues [2004] consider to be negligibly affected remodeling). Seventeen growth rings are present in the outer half of the compacta, while the rest of the compacta consists of dense Haversian bone (i.e., all growth rings formed in early ontogeny are completely obliterated). Indeed, a part of the bone wall is so extensively remodeled that not a single growth ring is evident. Horner and Padian (2004) also found considerable reconstruction in more than half of the compacta of a *T. rex* fibula and had to resort to complicated back-calculation methods to ascertain the number of growth rings removed as a result of remodeling. I am not convinced that remodeling in non-weight-bearing bones is as negligible as suggested by Erickson and colleagues (2004) and until this is more rigorously tested, I do not advocate the use of these skeletal elements for skeletochronology.

Not all dinosaurs have zonal bone in the compacta. Armand de Ricqlès and Robin Reid have documented uninterrupted fibrolamellar bone in the bones of several isolated bones of dinosaurs. In my analyses of a growth series of thirteen femora of *Dryosaurus lettowvorbecki* from Tanzania, I found that throughout ontogeny, the dinosaur formed bone in an uninterrupted manner (Chinsamy 1995b). I also found such tissue in the bones of several hypsilophodontids from the polar locality of Australia, Dinosaur Cove. Thus, it appears that dinosaur compact bone can consist of either interrupted (i.e., zonal bone) or uninterrupted fibrolamellar bone tissues. In an interesting study of sauropods from Tendaguru, Martin Sander (2000) found that the *Barosaurus* bones could be separated on the basis of different bone types (i.e., he found that five humeri, femora, and a fibula showed a type A bone microstructure, which was an uninterrupted azonal bone tissue type, while type B in a humerus, three femora, and a tibia had zonal bone tissue). Sander seems to think that these differences in histology reflect sexual dimorphism. I am not convinced because the sample size is relatively small and feel that perhaps they reflect taxonomic differences. Sander does explore this possibility but says that because the femora are morphologically similar, they rather reflect sexual dimorphism. However, in our study of the diagnostic remains of contemporary cynodonts, *Cynognathus* and *Diademodon*, which, except for distinctive characteristics in the skull are indistinguishable from one another on the basis of their postcranial skeletons, we found distinctive characteristics in their bone microstructure (i.e., the former having azonal bone, while the latter consistently had zonal bone; Botha and Chinsamy 2000).

Another characteristic of the Tendaguru sauropods is, except for *Barosaurus* as described above, most of the bones show a scarcity or absence of LAGs and no annuli or even cyclical modulations in the rate at which fibrolamellar bone formed (Sander 2000). In addition, the compact bone in these sauropods appeared to be dominated by

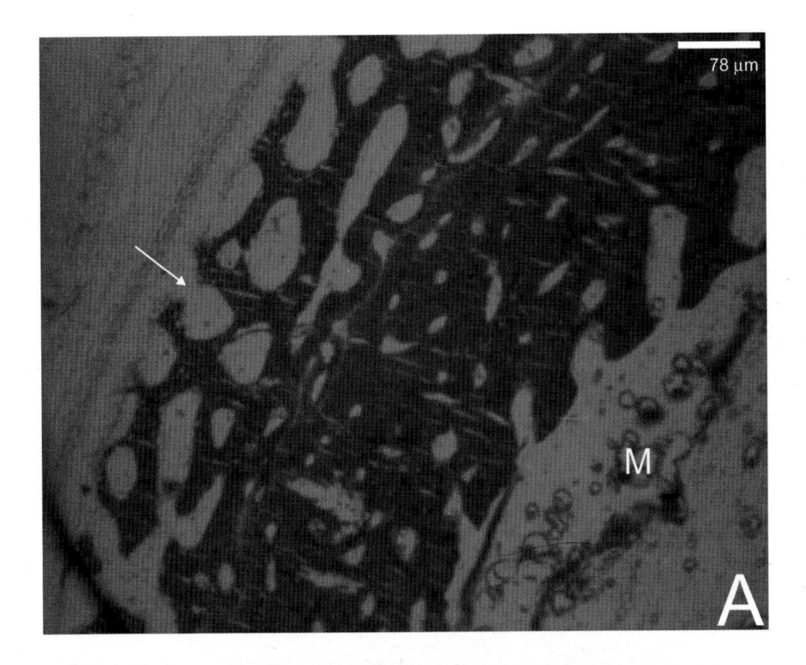

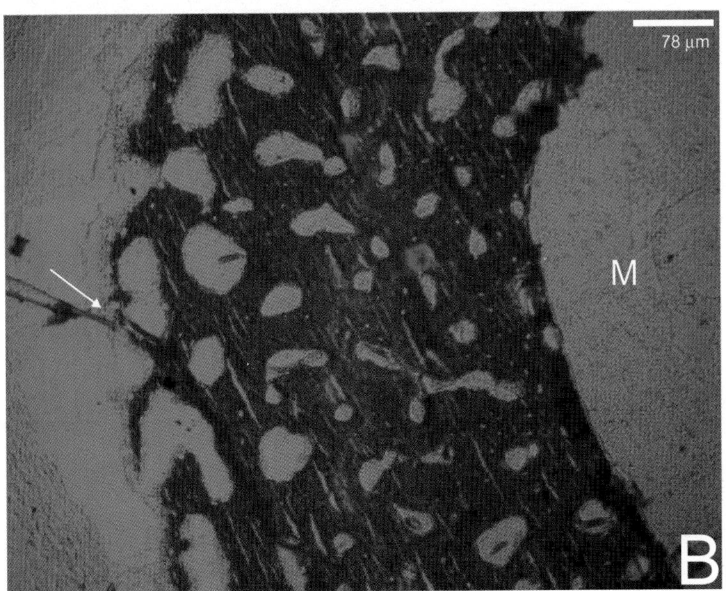

Plate 1. Fluorochrome labeling in bones of the Japanese quail (*Coturnix japonica*). The red staining indicates the differential amounts and location of bone that formed. *A*. A femoral cross section of a bird that was kept on a restricted diet before the labeling. *B*. The comparatively slower rate of bone formation in a radius. *M* indicates the medullary cavity and the arrow indicates the periosteal (outside) surface of the bone. Images courtesy of Matthias Starck.

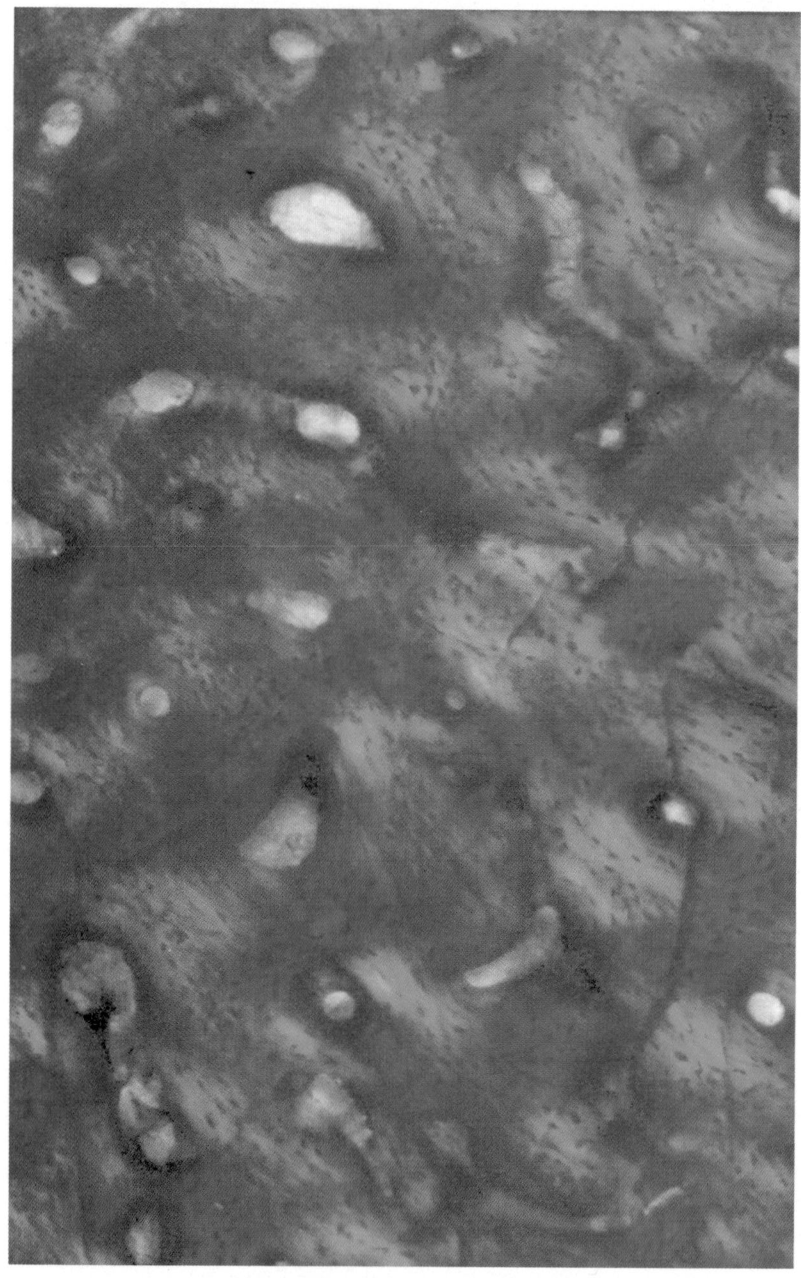

Plate 2. Early stages of fibrolamellar bone formation in a femur of *Dryosaurus altus*.

Plate 3. Longitudinal section of a femur of *Dryosaurus lettowvorbecki* showing calcified cartilage.

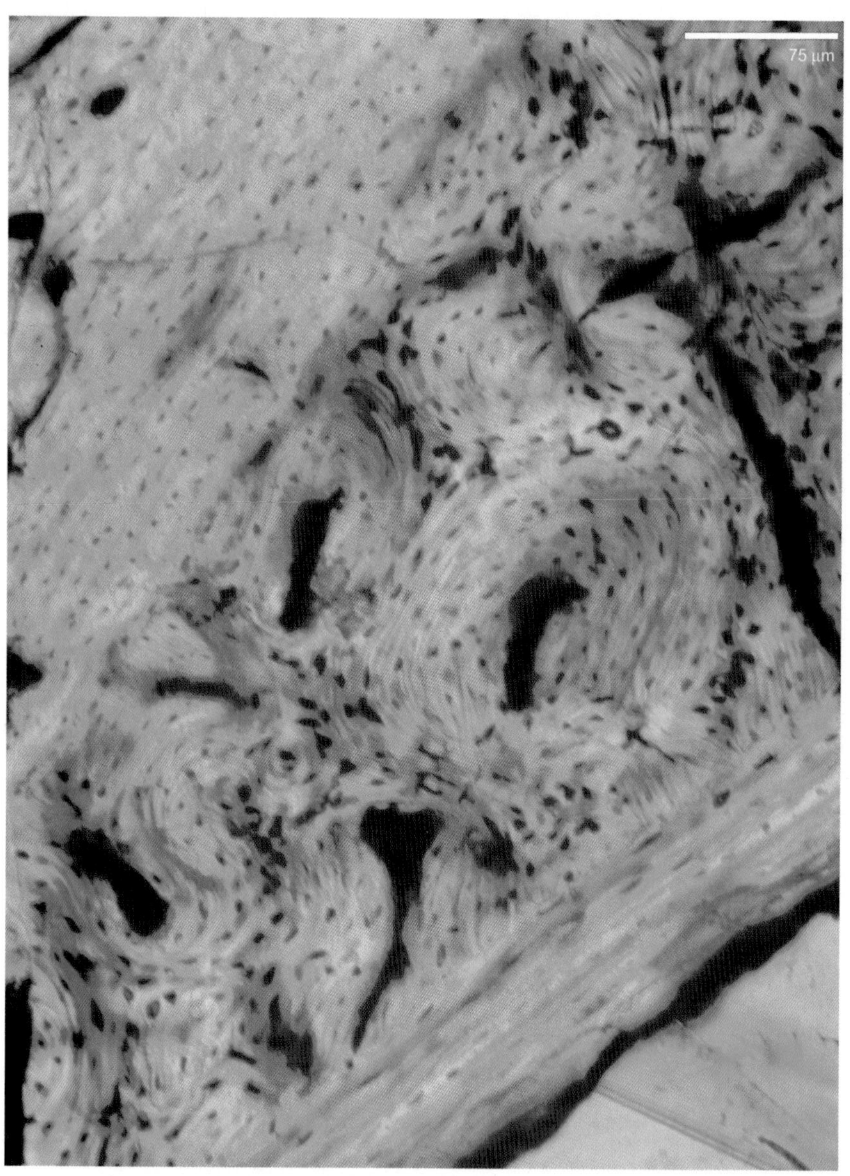

Plate 4. Section of compacted coarse cancellous bone sandwiched by primary periosteal fibro-lamellar bone on the peripheral side and by endosteally formed inner circumferential lamellar bone in the perimedullary region.

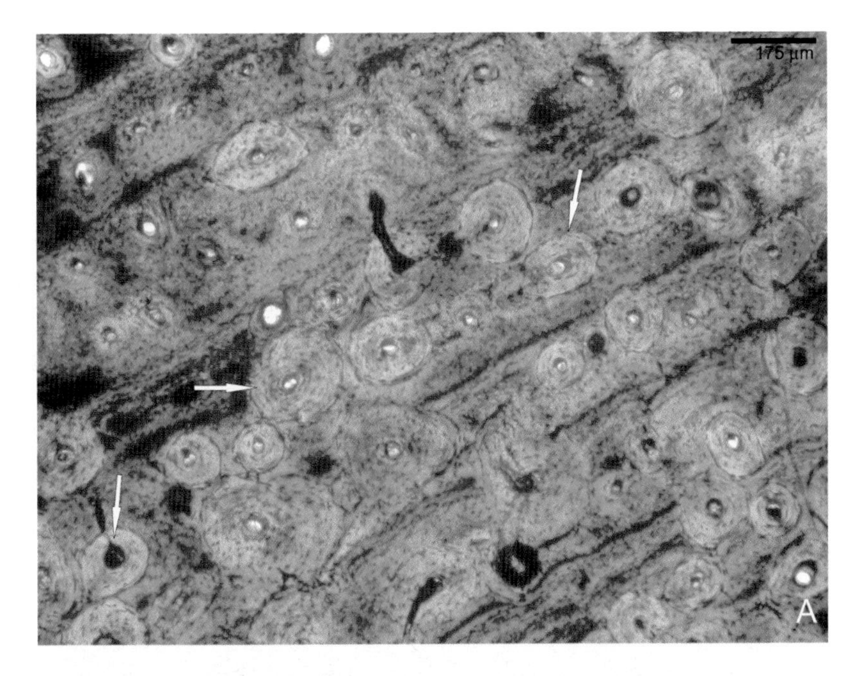

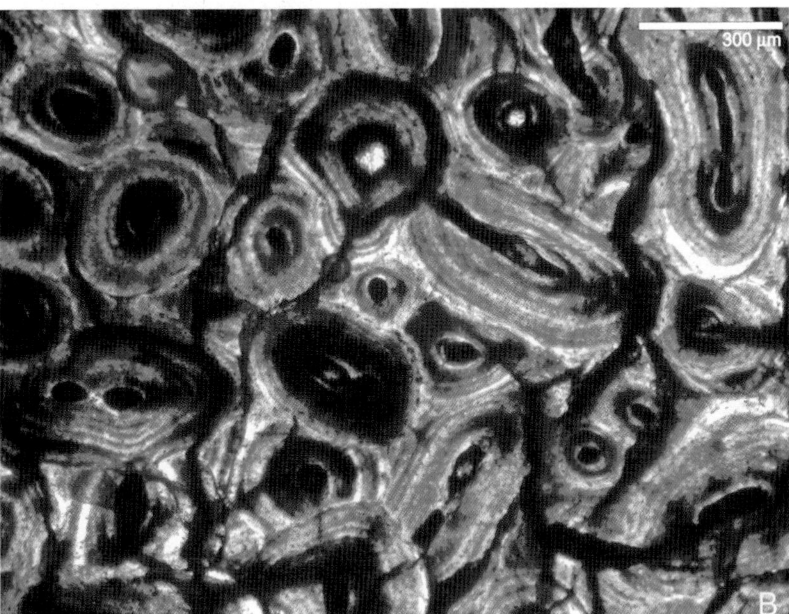

Plate 5. *A*. Zonal bone in a *Tyrannosaurus rex* fibula. Although a number of secondary osteons (arrows) are present in the compacta, the primary bone tissue is still visible, and distinctive growth lines are evident. *B*. In this sauropod bone from Niger, the bone consists entirely of dense Haversian bone. Note that the bone in-between the secondary osteons are also remnants of secondary osteons.

Plate 6. *Patagopteryx* femur showing the highly vascularized compacta and a line of arrested growth (arrow).

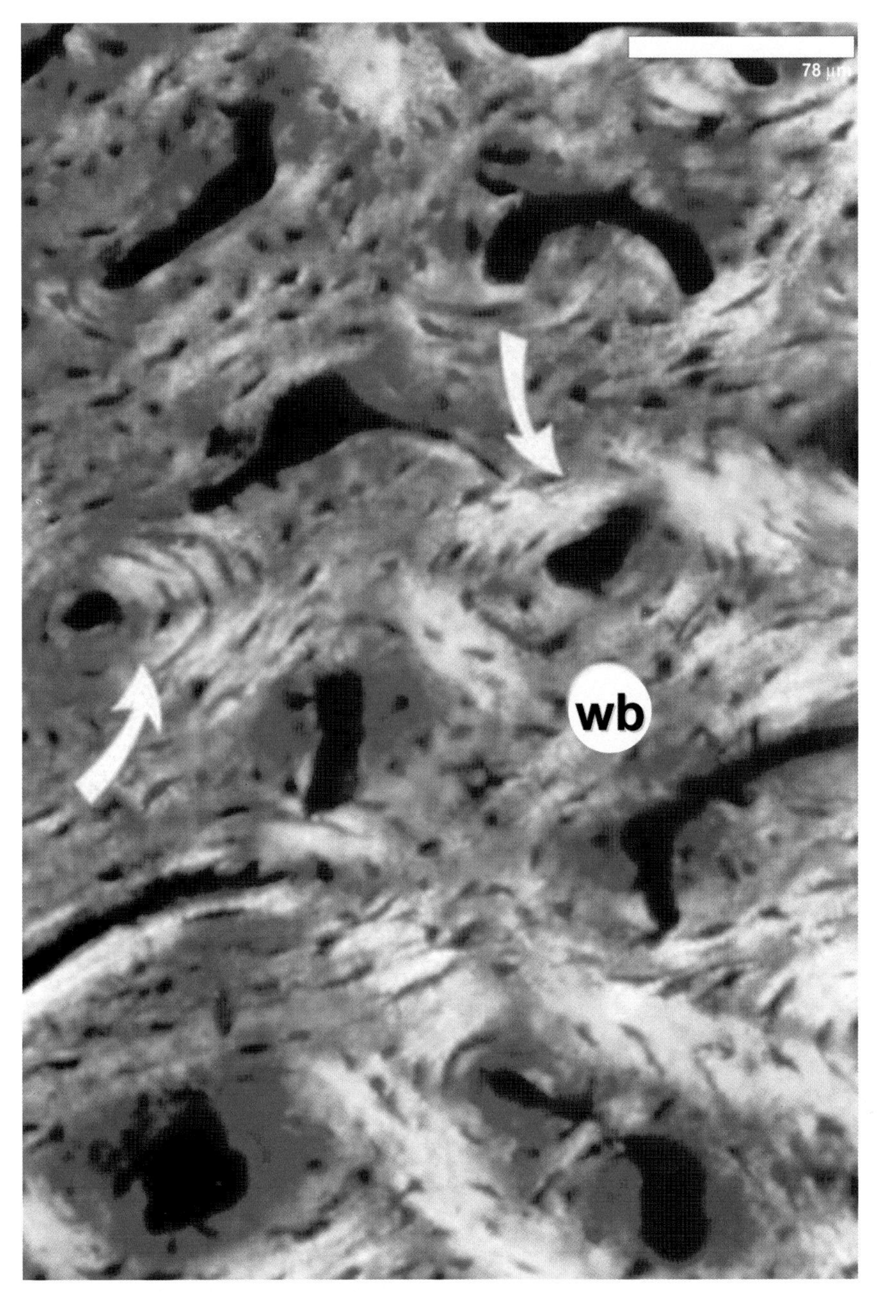

Plate 7. Fibrolamellar bone in a 190-million-year-old *Syntarsus rhodesiensis*. Arrows indicate primary osteons; *wb*, woven bone texture of the bone matrix.

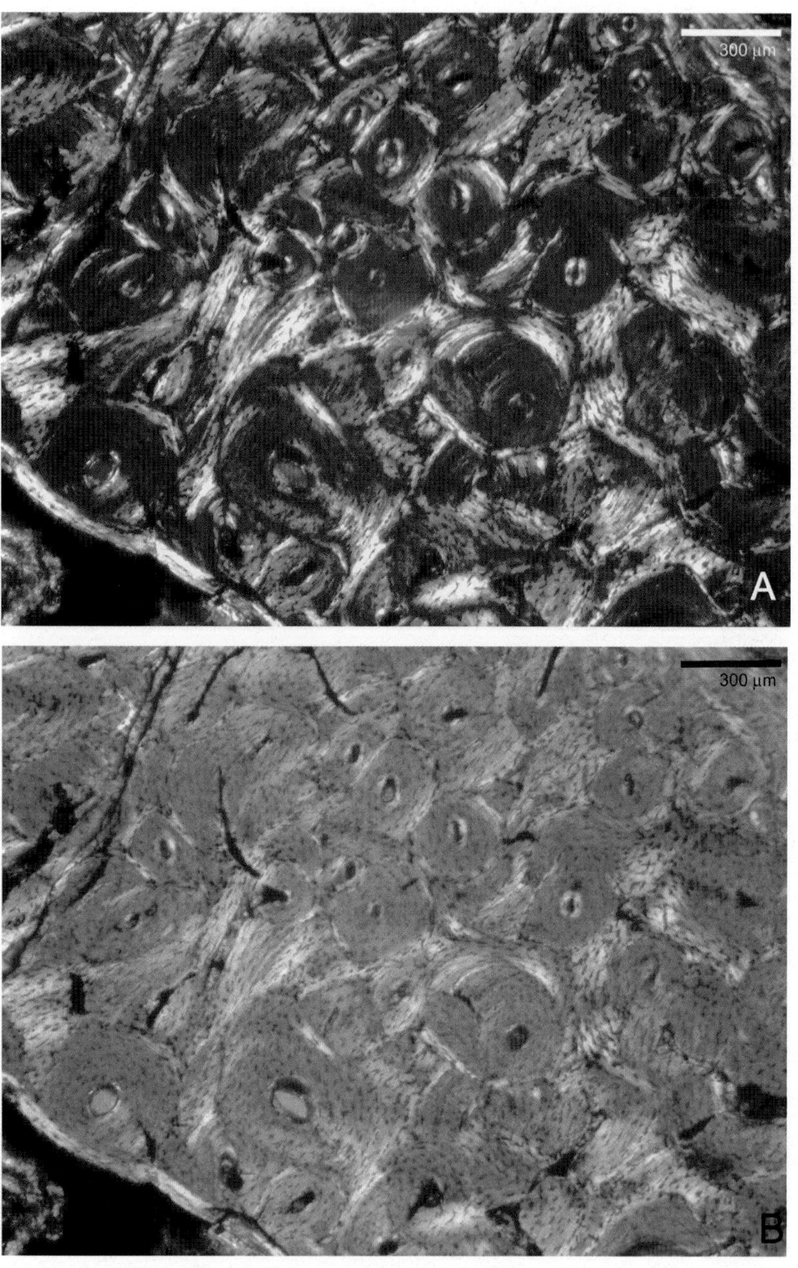

Plate 8. Haversian bone in *Archaeornithomimus*, under polarized light (*A*) and with a quarter lambda filter (*B*). The different colors indicate the "fibrillar" orientations.

primary fibrolamellar bone and, in general, exhibited a low degree of remodeling—this is particularly true for *Dicraeosaurus* and the type A *Barosaurus* bones. However, several other sauropods apparently do have zonal bone in their compacta, for example, sauropod pubis, *Camarasaurus* chevron bone (Reid 1990), and *Bothriospondylus* humerus (de Ricqlès 1983).

In her study of *Apatosaurus,* Kristi Currey-Rogers (1999) found that although the scapula lacked typical zonal bone, local variation in bone depositional rate was reflected in the vascularization of the compact bone. Currey-Rogers describes areas of low vascularity with mainly longitudinally oriented osteons and some reticular anastomoses, alternating with wider areas of richly vascularized reticular and laminar tissues. In our study of Japanese quail, Matthias Starck and I documented numerous small, longitudinally oriented vascular canals in individuals that were given a restricted diet as compared with those that were able to feed freely (Starck and Chinsamy 2002). Given our findings, it is possible that the cycles observed in *Apatosaurus* scapulae correspond to periods of nutritive stress, which could be seasonally linked.

Outer Circumferential Layer

This slowly formed periosteal bone is formed in the outermost part of the bone of animals that have reached their mature body size. As mentioned in chapter 3, this outer circumferential layer (OCL), which can consist of either lamellar or lamellated bone, is formed in animals that have a determinate growth strategy, such as mammals (Figs. 3.4, 3.9) and birds (Fig. 3.11). In general, OCL tends to be avascular and may or may not have periodic rest lines interrupting the slow rate of bone deposition.

Such bone tissues have been identified in several dinosaur taxa, for example, *Brachiosaurus* (Gross 1934); *Iguanodon* and *Rhabdodon* (Reid 1984a, 1990); a Transylvanian dinosaur (Campbell 1966); *Janenschia* (Sander 2000); a polar hypsilophodontid (Chinsamy et al. 1998); as well as in the theropods, *Syntarsus* (Chinsamy 1990); *Troodon* (Varricchio 1993); *Saurornitholestes* (Reid 1990); and *Allosaurus* (Reid 1996). In *Ceratosaurus,* Reid (1996) described lamellated bone occurring in the peripheral region of a limb bone at the end of growth.

However, many dinosaurs do not show such slowly accreted bone in the peripheral regions of their compacta. This seems to support de Ricqlès' (1980) suggestion that dinosaurs grew throughout their lives (i.e., experienced indeterminate growth). In a growth series of femora of the prosauropod *Massospondylus carinatus* (Chinsamy 1993a), and in the ornithopod *Dryosaurus lettowvorbecki* (Chinsamy 1995b), I have not been able to identify such accretionary bone even in the largest bones of the sample and have suggested that these animals are still growing. However, the possibility exists that perhaps the largest individuals have not been sampled.

Although the OCL refers only to appositional (i.e., diametric) growth, its presence or absence has been used to deduce determinate and indeterminate growth strategies. To check whether this corresponds with a slow down in longitudinal growth, longitudinal sections of the distal/proximal ends of the bone need to also be studied.

Haversian Bone

It is well documented that dinosaur bone often shows a large amount of secondary reconstruction (Haversian reconstruction), which leads to the formation of secondary osteons (Plate 5A). In many instances, Haversian reconstruction can continue to such an extent that bone between secondary osteons is itself secondary in nature and results in a tissue that is referred to as *dense Haversian bone* (Plate 5B). This type of tissue has been observed as particularly well developed in bones of sauropods. However, many dinosaurs do not show such extensive development of Haversian bone. A tibia and fibula figured by Madsen (1976) show a compacta consisting of mainly of primary bone tissues. *Coelophysis* (A. Chinsamy-Turan, personal observation) also has a lower amount of Haversian reconstruction as compared with a similar-sized *Syntarsus rhodesiensis* from Zimbabwe (Chinsamy 1990). Why such differences occur is uncertain. It may be a result of differences in the level of phosphocalcic metabolism or, perhaps, even a response to biomechanical stress.

As is the case among modern vertebrates, secondary reconstruction tends to increase with age. However, it is interesting to note that, in the smallest femur (about 31% of the largest adult size) of the growth series of *Dryosaurus* that I studied, isolated and completely formed secondary osteons were already present in the compacta (Chinsamy 1995b). This is indicative that secondary reconstruction began very early in the life of this medium-sized ornithopod dinosaur. This characteristic early onset of secondary reconstruction is generally typical of mammalian bones (e.g., it can begin after seven weeks in a dog and by eight months in a human baby; Lacroix 1971).

Secondary osteons also tend to occur in the innermost parts of the cortical bone, nearer the medullary cavity. The predominant location of mechanically "weaker" secondary osteons toward the medullary region is perhaps important in terms of biomechanical functioning of the femur: primary, unremodeled bone is more fatigue resistant and stronger in tensile tests to failure than Haversian bone (Carter and Spengler 1978) and therefore having this bone near the periosteal surface enhances the strength of the bone because during life, higher stresses are distributed furthest from the longitudinal neutral axis.

Endosteal Compact Bone

MEDULLARY LINING BONE (INNER CIRCUMFERENTIAL LAYER). Following medullary enlargement, growth reversal occurs, and the endosteal deposits of the poorly vascularized, slowly formed lamellar or lamellated bone occurs around the medullary cavity (e.g., in birds, see Fig. 3.11). Reid (1996) refers to this tissue as *medullary lining bone*, while Enlow and Brown (1957) use the term *inner circumferential lamellae*. Because this bone can be either lamellar or lamellated, it is better to use the term *inner circumferential layer* or *medullary lining bone*. In several dinosaurs, an inner circumferential layer (ICL) of bone is observed lining the medullary cavity (e.g., *Iguanodon*, *Rhabdodon*, *Timimus* (Fig. 3.3C) and *Syntarsus* (Fig. 5.3A). The external boundary of the endosteal bone is usually distinctive because of the presence of a tide line that marks the extent of bone resorption from the medullary surface. Sometimes the ICL itself may show resorption, followed by redeposition of bone.

COMPACTED COARSE CANCELLOUS BONE. This type of bone involves the endosteal infilling of the cancellous spaces (i.e., intertrabecular spaces with lamellar bone). The occurrence of such bone indicates that the area was once located near the metaphysis at an earlier stage of growth. Reid (1996) notes that such bone is not often seen in dinosaur bone and suggests that perhaps compaction and drifting were not important in the growth processes of dinosaur bones. However, I have observed compacted coarse cancellous bone in several distally and proximally located sections of *Dryosaurus* femora (Plate 4). In fact, in some of the sections, virtually the entire compact bone wall (from the periphery to medullary cavity) consisted of consisted coarse cancellous bone (Chinsamy 1995b). The occurrence of this tissue indicates that this region was once part of the metaphysis and now because of growth and remodeling processes, has become relocated into the diaphysis of the growing bone.

COMPACTED FINE CANCELLOUS BONE. Fine cancellous bone (i.e., produced by endochondral replacement of cartilage near the epiphyseal plate) can also be compacted to result in compact bone. Spicules of calcified cartilage are incorporated into the compact bone tissue (Enlow 1963). I have not observed the latter in bones of dinosaurs, although I have seen such bone in the proximal regions of the Mesozoic mammal, *Barunlestes*, from

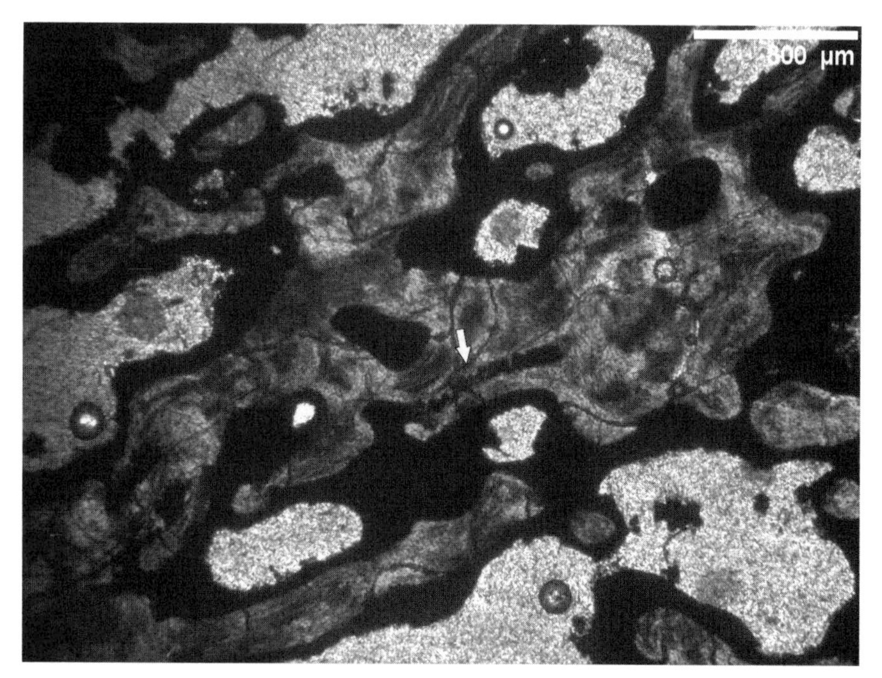

Fig. 4.5. Endochondral bone in a hadrosaur. The arrow indicates a small patch of calcified cartilage.

the Gobi Desert, and Enlow (1963) mentions that it is commonly found in the cortical bone of white rats and, in general, occurs in the bones of rapidly growing individuals.

From the above, it is apparent that a host of different types of periosteally and endosteally formed compact bone tissue can occur in different parts of the skeleton, in different parts of a single skeletal element, and even in a cross section of a single bone. Even in bones of similar tissue types subtle variations can occur, for example, an increase in the number of Sharpey's fibers in the region of a site of muscular attachment. Exactly what triggers the formation of different bone tissues is complex and appears to depend on a number of genetic and epigenetic factors.

Cancellous Bone

Endochondral Bone

Cancellous bone can be formed by endochondral ossification that involves the replacement of cartilage (Fig. 4.5). Such tissues are well documented in dinosaurs, for example, Reid (1984) has described such bone in the centra of *Iguanodon,* and in the epiphyses of

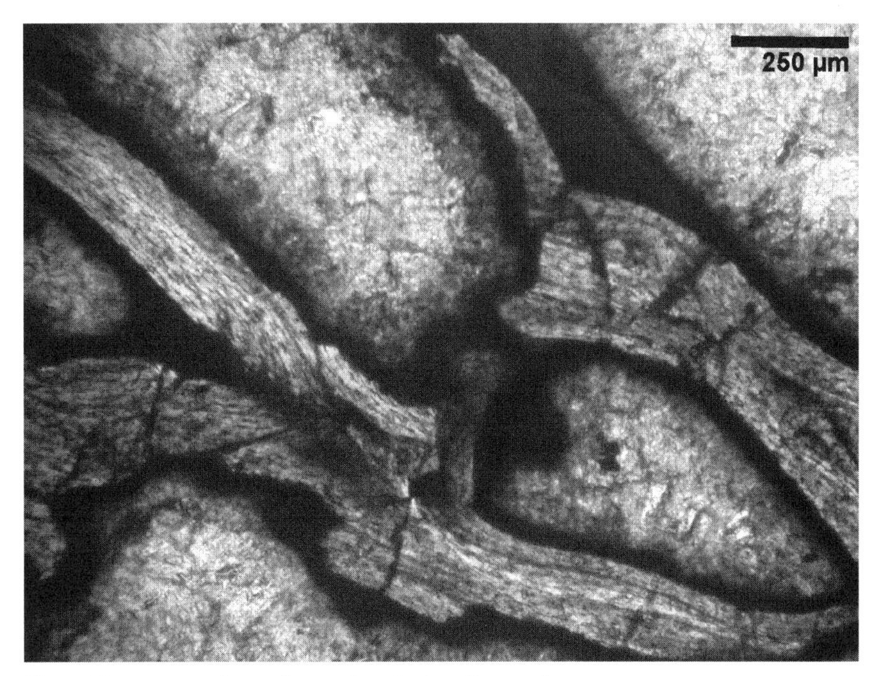

Fig. 4.6. Reconstructed cancellous trabeculae in a dinosaur bone.

Iguanodon, Hypsilophodon, and *Valdosaurus* limb bones; I have observed such bone tissue in the epiphyseal regions of *Dryosaurus* (Chinsamy 1995b) and in *Nqwebasaurus* femora; and Horner et al. (2000a) have described it in *Maiasaura.* In many adult dinosaurs, islands of calcified cartilage are visible within the endochrondrally formed trabeculae (Fig. 4.5). This characteristic suggests that endochondral bone formation occurred at a rapid rate.

Secondary Cancellous Bone

As the name suggests, this tissue is formed by resorption of either primary cancellous bone or primary compact bone. In the former case, bone is resorbed from the walls of the trabeculae, and subsequent redeposition of lamellar bone occurs along the trabecular walls. The overall texture of the bone can still remain cancellous. Such reconstructed cancellous bone is common in dinosaur bone tissue and usually forms most of their cancellous bone except in the metaphysis (Reid 1996). Indeed, most dinosaur vertebral cancellous bone is secondary cancellous bone tissue (Reid 1996). The bone is easily recognizable because tide lines showing the extent of bone resorption are discernible in the trabeculae (Fig. 4.6).

The resorption of primary compact bone to form cancellous bone is easily recognizable. It usually involves resorption of bone from around the vascular channels without subsequent redeposition, which results in the former compact bone being transformed into cancellous bone tissue. This type of bone is well described from dinosaur bone tissues.

Other Types of Bone

Metaplastic bone

Metaplastic bone is produced by the ossification of tendons and ligaments and has been described in dinosaurs since the late 1800s, for example, in *Iguanodon* (Dollo 1886) and in *"Anatosaurus"* (Broili 1922; Moodie 1928). According to R. Wheeler Haines (1969) the tendinous tissue becomes mineralized to form metaplastic bone in which the fibers and cells are preserved. Osteoderms are usually considered as metaplastic in origin, but they are dealt with separately because recent hypotheses suggest that they may not be metaplastic.

Osteoderms

A range of different bone tissues has been described in dinosaur osteoderms. The structure is highly complex and is variable even in a single dinosaur, depending on its location on the body. For example, the osteoderm structure of a stegosaur dorsal plate is distinctly different from that of an ossicle located laterally on the body of the stegosaur. In a study of the dorsal plates of *Stegosaurus,* Vivienne de Buffrénil, Jim Farlow, and Armand de Ricqlès (1986) found that the bone histology was highly variable from the base of the plate to the apex. Overall the plate appeared to be unexpectedly light and hollow. The peripheral parts of the plate consisted of thin sheets of compact bone tissue, while the central part consisted mainly of cancellous bone tissue, with thin bony trabeculae forming irregular medullary spaces. The primary cortical bone consists of sparsely vascularized lamellar-zonal bone, while in the deep part of the compacta, Haversian reconstruction is extensively developed. Following this region, the Haversian bone itself is cored by large cancellous spaces that have a complex network of trabeculae that also show reconstruction. The lower third of the plate consist of abundant Sharpey's fibers in mainly a primary cortical bone tissue. Distinct patterns in the orientation of the fibers (i.e., symmetrically arranged on either side of the sagittal plane of the plate) and biomechanical considerations led the researchers to propose a vertical orientation of the plates along the dorsal midline of the body (de Buffrénil et al. 1986). They further suggest that the plates could not have functioned as armor in the

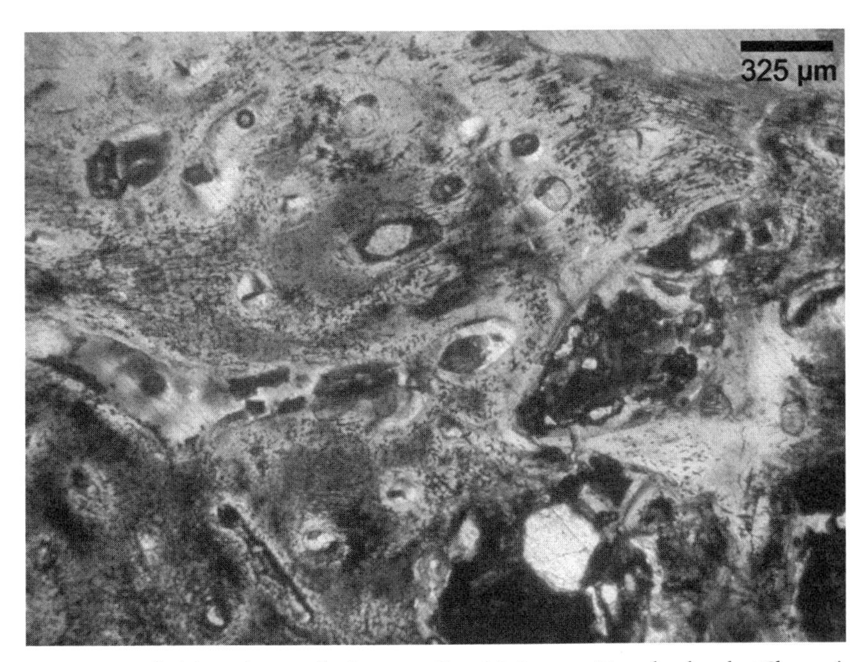

Fig. 4.7. An ossified dermal scute of a titanosaur from Madagascar. Note the abundant Sharpey's fibers in the bone.

usual way but that given the intricate nature of its architecture it was more likely that they functioned in thermoregulation.

In a study of the microstructure of ankylosaur dermal ossifications from the Antarctic Late Cretaceous, Armand de Ricqlès and his Argentinian colleagues (de Ricqlès, Pereda Suberbiola et al. 2001) found that the small, buttonlike ossicles have densely packed parallel fiber bundles, abundant Sharpey's fibers, and poor vascularization. Although they appear to be metaplastic in origin, conflicting structural characteristics were found: the fibrillary system appears to be arranged according to a radial growth pattern, and the mesh size and diameter of the bundle fibers appear to also increase radially. This would not have been the case if they were formed by mineralization of existing dermal tissue. The researchers therefore suggest that the ossicles could have formed *de novo* (i.e., as original elements through neoplasia) from radiation centers located in the dermis (at the limit of the stratum spongiosum and stratum compactum).

In his study of the dermal plates of titanosaurs, Argentinian paleontologist Leonardo Salgado (2003) has suggested that perhaps the bony plates are not osteoderms but, rather, true bones originating from the epiphyseal tip of the neural spines. Salgado based his hypothesis on Haines's findings (1969) that the epiphysis of the neural spines of reptiles

Fig. 4.8. Pathological bone from a healed fracture of a hadrosaurid neural spine. The remodeled fracture callus overlies the original periosteal surface (arrows). Image courtesy of Allison Tumarkin-Deratzian.

(e.g., *Sphenodon)* generally remain cartilagenous and that, on occasion, in this region additional ossifications can occur.

Pathological Bone

Depending on the nature of the pathology this bone could be compact or cancellous. Pathological bone has been recognized in dinosaurs since the early 1900s (e.g., Moodie 1923), and more recently, Bruce Rothschild and various colleagues have documented a host of pathologies in dinosaur bone (e.g., Rothschild and Martin 1993; Rothschild et al. 1997).

Diffuse idiopathic skeletal hyperostosis (DISH) is a condition in which ligaments and tendons calcify at their points of attachment to bone. DISH appears to have a high frequency among dinosaurs and has been recognized in ceratopsians, hadrosaurs, iguanodontians, and pachycephalosaurs along the spinal longitudinal ligaments, which are parallel to the long axis of the vertebral column and serve to keep it aligned. It is possible that the ligamentous ossification in these groups allowed the tail to be kept high off the ground and stabilized lateral movement. Stegosaurs probably used their tail for defense, and they do not show any DISH. Rothschild and paleontologist David Berman (1991) have observed that DISH has an incidence of 50% in *Diplodocus* and 25% in *Camarasaurus* specimens. The affliction usually involves fusion

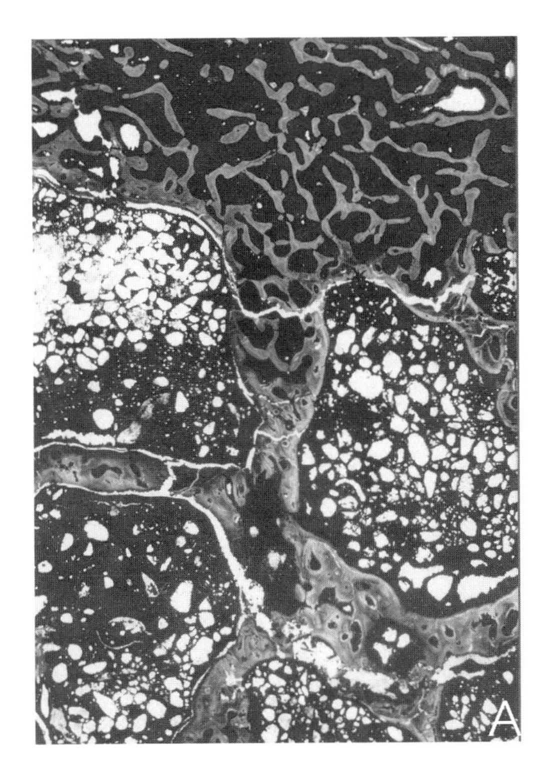

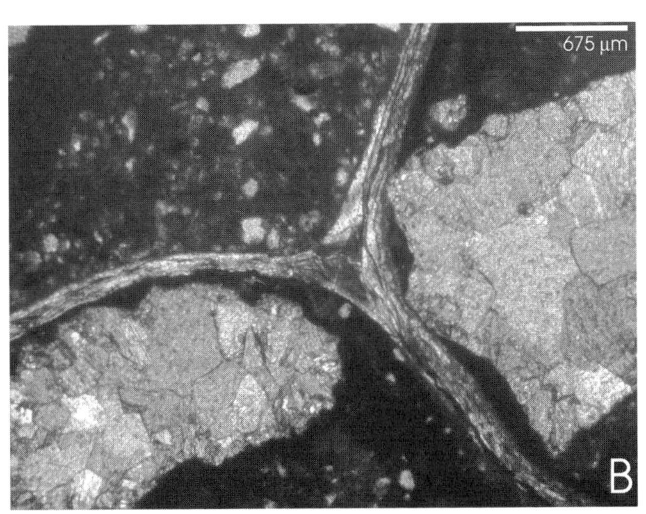

Fig. 4.9. Cavernous bone in a sauropod vertebrae. *A*. Normal cancellous bone is resorbed to form large spaces typical of cavernous bones. The surfaces usually become lined with lamellar bone. Image courtesy of Robin Reid. *B*. Extremely thin partitions of cavernous bone of a sauropod vertebrae.

of two to four vertebrae in caudal positions 17–23, which would have stiffened the tail of the animal. The researchers propose that this would maintain the cloaca in a position more accessible for mating and further suggested that DISH in these sauropods identifies males of the species.

Although pathological bone has been reported, these are usually described through x-ray analyses and morphological descriptions, without microstructural details. One of the few instances wherein this has been provided is in Campbell's (1965) description of a bone lesion in a dinosaur bone. In this paper, he described the hyperostotic periosteal outgrowths from the bone surface (referred to these days as *reactive bone*), having radial honeycomb-like structures, which have some degree of intercommunication.

Osteoporotic bone in which the primary bone has been largely resorbed and not completely replaced by Haversian bone has been described in an *Apatosaurus* scapula and a *Camarasaurus* limb bone.

I have rarely observed pathological bone microstructure in dinosaurs. My personal experience is only of the healed fracture callus of a hadrosaur's dorsal neural spine from Alberta (Fig. 4.8). Darren Tanke of the Royal Tyrrell Museum in Drumheller, Canada, has found that about 1% of ceratopsians and hadrosaurs from Alberta showed fractures in the midposterior dorsal ribs. Among the iguanodontians, such as *Camptosaurus* and *Iguanodon,* the fractures appeared to be on the dorsal neural spines. Rothschild and Tanke (1991) have suggested that these fractures were caused by the weight of the male on the female's back during copulation.

Cavernous Bones

Cavernous bones contain large internal spaces (several centimeters wide) between plates of compact bone (Reid 1996; Fig. 4.9). Cavernous bones are known from bones of birds into which the air sac system extends and are also found in the skulls of elephants (Reid 1996). Cavernous ribs and vertebrae have been recognized in bones of theropods and sauropods. Reid (1996) described the cavernous spaces in vertebrae of sauropods as having been formed by large-scale reconstruction processes. The nature of the compact bone is quite variable and could be primary periosteal bone, Haversian bone, endochondral bone, lamellar/lamellated bone of the ICL, or medullary lining bone (Reid 1997). It is obvious that such cavernous bone would have lightened the mass of the vertebrae, and as early as 1947, Janensch suggested that they may have been air filled (pneumatic). Using the extant phylogenetic bracket (EPB) approach, Matthew Wedel (2003) has suggested that the complex architecture of sauropod vertebrae with its laminae, fossae, and complex cavenous

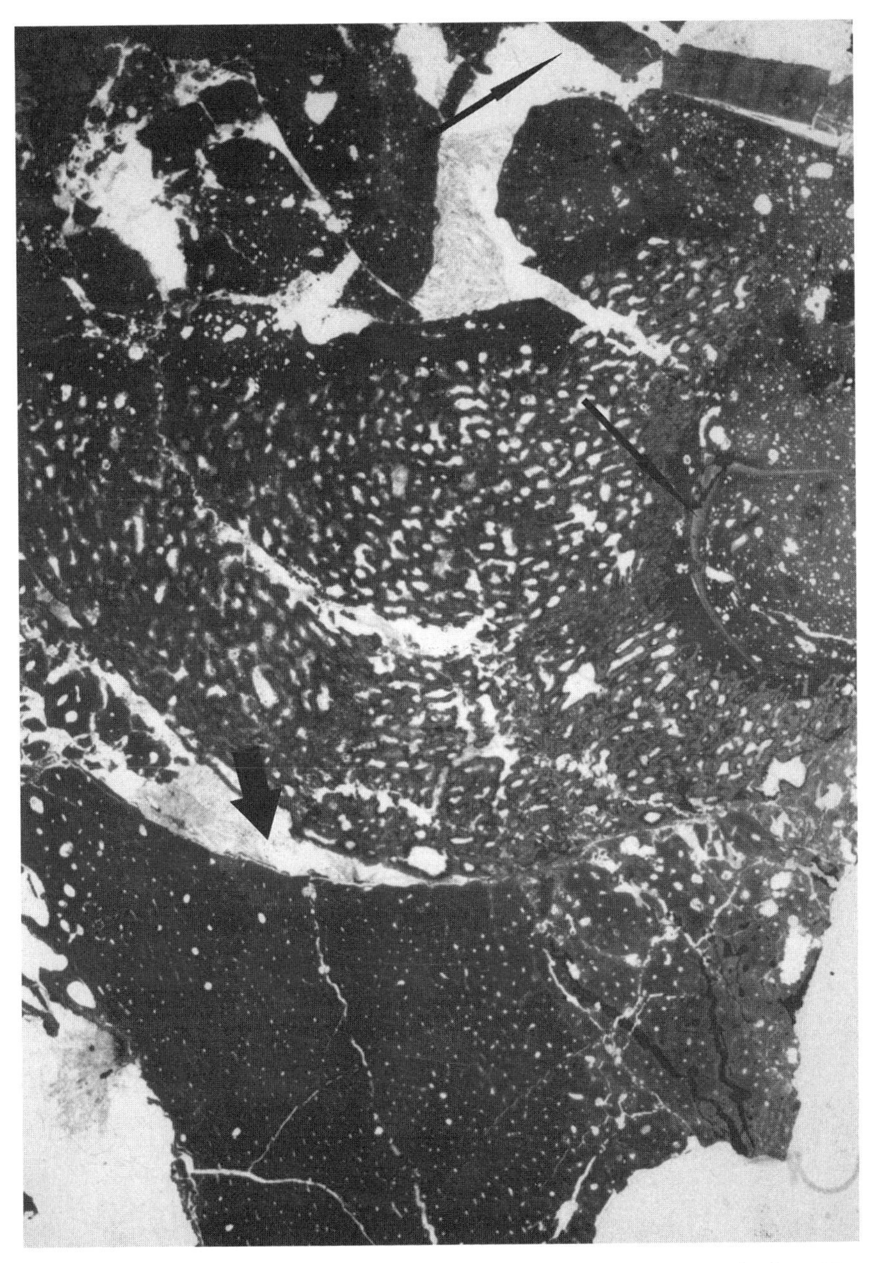

Fig. 4.10. Cancellous bone of attachment and Haversian bone of the dentary are clearly visible. Thick arrow points to the dentary. Thin arrows point to remains of teeth in the bone of attachment. Image courtesy of Robin Reid.

bone structure are osteological correlates for a system of air sacs (Wedel 2003) as seen in birds.

Dental Bone (Alveolar Bone or Bone of Attachment)

Dinosaur teeth, like basal archosauromorphs and crocodilians, are formed in deep grooves in tooth-bearing bones in which the sockets consist of alveolar bone. Such bone tissue has been recognized in bones of *Allosaurus, Camarasaurus, Camptosaurus,* and *Marshosaurus* (Reid 1996; Fig. 4.10).

Conclusions

The microscopic structure of dinosaur bones is generally well preserved, and today we have a reasonable understanding of the types of bone tissues that make up dinosaur bone. We know that different parts of the skeleton can have different bone tissues, and we see that different regions of a single bone and even within a single cross section of a bone can have different microstructure. Several different types of compact and cancellous bone tissues are recognized in dinosaur bones, and the occurrence of each of the tissues provides information about the qualitative rate at which the bone was formed, as well as how the bone formed. For example, the occurrence of zonal bone directly indicates that the animal experienced alternating periods of slow and fast rates of growth, while the presence of Haversian bone implies that bone resorption of an earlier primary bone must have occurred. By comparing dinosaur bone tissues with that of extant animals (particularly with those representing its EPB, as well as morphological and ecological equivalents) biological implications of dinosaurian bone microstructure can be deduced.

CHAPTER FIVE

Growing Dinosaurs

As with any other animal, conception marks the beginning of the life history of a dinosaur and death marks the end. Embryonic development, growth in size, the acquisition of morphological characteristics until reproductive age (i.e., maturation), and senescence lie between these end points. Such growth stages are readily observable among extant populations. However, among fossil animals, the different growth stages, particularly the developmental stages during late maturation, are rather difficult to discern.

Although we have absolutely no idea of pre–egg laying embryonic development in dinosaurs, considering the extant phylogenetic bracket of dinosaurs, we can infer that dinosaur development followed similar patterns to their closest living relatives (i.e., birds and crocodilians). With this in mind, we can deduce that during early embryonic growth, the single-celled zygote is cleaved into a multicellular stage. In many extant vertebrates, the way cleavage occurs is variable and depends on the amount of yolk in the egg. Considering that both birds and crocodiles produce eggs with large amounts of yolk, it is reasonable to deduce that dinosaur eggs also contained large yolk sacs with protein-rich food for the developing embryo. Therefore, cleavage would probably have been similar to their extant relatives, and the resulting blastomeres would have clumped at the animal pole of the egg to form the blastoderm. Gastrulation and neurulation would have then followed, and at the end of neurulation, the major accomplishments of the dinosaur embryo would be the attainment of bilateral symmetry and the development of the three germ layers, namely, ectoderm, endoderm, and mesoderm. The ectoderm would eventually form the nervous tissue and the epidermis, the endoderm would form the digestive and respiratory systems, and the mesoderm would form the connective tissues (including the skeleton) and muscular and circulatory systems.

Bird eggs are generally laid when the embryo is at an early stage of growth (at about the late blastula stage), while crocodilian eggs are laid when the embryo is at a more advanced stage of development (at about midneurulation stage). Regarding dinosaurs, we are uncertain whether eggs were laid with a fairly advanced embryo, as in crocodilians, or with a much more immature embryo, which is the case for many extant birds.

Dinosaur Eggs, Nests, and Embryos

Abundant dinosaur eggs, depicting a wide array of shapes and sizes, some in nests and some isolated, are known from various parts of the world, and they provide a host of information regarding dinosaur reproductive behavior.

From the variety of nesting sites discovered, it appears that dinosaurs constructed a wide variety of nests ranging from simple holes in the ground to more complicated mounds. The hadrosaur, *Maiasaura peeblesorum* from the Upper Cretaceous Two Medicine Formation in Montana, is probably the best-known dinosaur exhibiting a communal nesting strategy. Jack Horner of the Museum of the Rockies in Montana described an assemblage comprising eight nests that were approximately 7 m apart. This spacing coincides with the approximate length of an adult *Maiasaura,* and Horner used this as evidence that adults attended the nests. A sauropod (titanosaur) communal nesting ground is also known from the Upper Cretaceous of northwestern Patagonia (Chiappe et al. 1998), where thousands of eggs, roughly 13–15 cm in diameter, were found in an area of more than 1 km^2. Yet another communal nesting site consisting of eight nests about 3 m apart from each other was described by David Varricchio of the Museum of the Rockies and colleagues (1997). These nests were initially thought to be that of the hypsilophodontid, *Orodromeus;* however, they are now considered to be those of the small theropod *Troodon*. In the analysis of these egg clutches, David Varricchio and colleagues (1997) found that the eggs were laid in pairs, which they interpreted as evidence of two functional ovaries and oviducts in the animal. Such anatomical details for dinosaurs were unknown at the time, but shortly thereafter *Sinosauropteryx* was described with a pair of eggs clearly visible within the body cavity (Chen et al. 1998). This verified Varricchio and colleagues' hypothesis of paired ovulation in *Troodon*. It is interesting that paired ovulation is a reptilian characteristic (birds have only a single functional ovary and oviduct).

Considering the extant phylogenetic bracket of dinosaurs, the best modern analogues for comparing dinosaur nesting behavior are birds (especially, paleognaths and megapodes) and crocodilians, which all lay eggs on the ground, although the actual nest construction is variable. Paleognaths construct rather simple, shallow holes in the ground, while, depending on the species, megapodes lay their eggs in simple holes in the ground or in burrows or complex mounds. Some species of crocodilians and megapodes also use communal nesting grounds. Once eggs are laid, they need to be incubated at a correct temperature for embryonic growth, and in the case of crocodilians, the incubating temperature determines the sex of the hatchlings. Crocodilians, irrespective of whether they are hole-nesting or mound-nesting species, use environmental heat (sand or vegetative litter) to incubate their eggs. Paleognaths use body heat, while megapodes are unique among birds because they (like crocodilians) use naturally occurring heat for incubation: microbial decomposition of leaf litter, geothermal activity, and solar radiation. Unlike crocodilians, which stay in the vicinity of the nests to protect and guard the eggs, megapodes generally abandon their eggs once they are laid.

In the 1995 American Museum of Natural History field season to the Gobi Desert, Mark Norell and colleagues (1995) made a stunning discovery: an *Oviraptor* "sitting" on a clutch of eggs. Shortly thereafter, Phil Currie, of the Tyrell Museum in Alberta and Chinese paleontologist Dong Zhiming of the Institute of Vertebrate Paleontology and Paleoanthropology in Beijing, also described a specimen found in 1990: an *Oviraptor* atop a nest of eggs. In 1997, when Varricchio and colleagues described the nests and egg clutches of *Troodon,* they too reported a partial skeleton (consisting of mainly hindlimb elements) of a *Troodon* in contact with a clutch of eggs. These three unique finds are interpreted as demonstrating incubating behavior in dinosaurs. Given the association of eggs and adult dinosaur remains, such an interpretation is indeed compelling. However, it is worth considering that although crocodiles do not apply body heat to their eggs, Walter Coombs (1990) has reported that they can lie on top of their nests when guarding their eggs. Is it possible then that instead of incubating the eggs the adult dinosaurs associated with the nests may be protecting or defending their nests (especially considering that the oviraptorosaurs atop their nests might have been caught in a windstorm)?

In the analysis of *Troodon* nests, Varricchio and collaborators proposed that the dinosaur eggs were partially buried and probably lacked a chalazae (a structure in avian eggs that supports the yolk and allows it to rotate freely, resulting in a permanently up position for the developing embryo). One of the important functions of incubating bird parents is to rotate the eggs so that heat is evenly distributed among the eggs. Thus, if dinosaurs lacked egg rotation, large clutch sizes would result in uneven heat distribution if the parent incubated the eggs with body heat, which would have impaired the development of the embryos. Given the wide variety of nest constructions evident among the Dinosauria, it is possible that different incubation methods were used. For example, according to Jack Horner and the late Bob Makela (1979), *Maiasaura* appears to have laid about eleven eggs in a bowl-shaped nest about 2 m in diameter and 0.75 m deep. In this case, incubation heat was probably obtained from decomposing vegetation.

The number of eggs deposited in the nests varied although it appeared that between twenty and thirty eggs per nest was typical. In the case of *Maiasaura,* eleven hatchlings were reported, and it is reasonable to assume that as many eggs were laid, unless some eggs did not reach term. Mark Norell and colleagues (1995) estimated twenty-two eggs in the *Oviraptor* clutch from the Gobi Desert, while Varricchio and co-workers (1997) suggested that *Troodon* laid about twenty-two to twenty-four eggs per nest. Such large numbers of eggs per nest suggests that they were r-strategists and could mean high mortality rates among the hatchlings and, as a result, low breeding success. Among extant birds, clutch size is variable and is dependent on a variety of factors such as age, body size, and nest type. A further factor is the degree of maturity of the hatchings. Two terms used in this context are *nidicolous* and *nidifugous* (Campbell and Lack 1985). The term *nidicolous* refers to young birds that remain in the nest after hatching and are not in all species psilopaedic (i.e., with little or no down when hatched) or even wholly altricial (helpless when hatched). The term *nidifugous* refers to young birds that leave the nest immediately or soon after hatching and are necessarily both precocial (active immediately after hatching) and ptilopaedic (clad in down when hatched). Because the energetic demands on parents of nidicolous hatchlings is extremely high, these birds tend to lay small clutches of eggs and in some instances only one egg at a time (e.g., Procellariiformes). Parents of nidifugous hatchlings tend to lay larger clutches of eggs. Thus, considering the clutch size of extant birds, the large clutches of dinosaurs suggest a low parental involvement after hatching because such a large number of altricial, nidicolous young would require an enormous amount of energy expenditure on the part of the parent(s).

Although more than 220 egg localities are known worldwide, only a small fraction are of complete nests with diagnostic embryos. Although embryonic dinosaurs have become relatively well known in the fossil record, it is extremely difficult to assign them to particular dinosaur taxa and therefore many of the embryonic dinosaurs are of unknown species. There are, however, some diagnostic embryos, such as *Troodon, Hypacrosaurus,* and *Oviraptor.*

Embryonic Age and Growth

An important consideration that needs to be emphasized is that most of the embryonic dinosaur skeletons known from the fossil record are of late-stage embryos. This is not unexpected because in earlier embryonic stages the skeleton would have largely been cartilaginous and therefore would have decomposed soon after death with other soft tissue and organic material inside the egg. Therefore, only if the embryonic cartilaginous skeletal precursor had already begun to ossify, could it have had any chance of becoming fossilized.

However, even with the limited numbers of known dinosaur embryos, it is clearly evident that several embryonic growth stages are present in the fossil record. Without having done an exhaustive survey, from my observations, it appears that the sauropod embryos from Auca Mahuevo in Argentina represent very early stages in embryonic development because shafts of long bones are mostly preserved (i.e., "articular" ends of the bones are not preserved). Given the way the bones grow (see Fig. 3.2), it is apparent that in these embryos only the initial stage of diaphyseal ossification had occurred when they died. However, it looks as though the *Hypacrosaurus* embryos represent more advanced stages of embryogenesis because the complete long bone (i.e., including "articular" ends) is preserved and, in thin section, calcified cartilage and endochondral bone is evident.

The embryonic bone from Alberta that I studied (Fig. 3.9) seemed to suggest an older age than the embryonic *Hypacrosaurus* described by Horner and Currie (1994) because, in this case, the woven bone matrix is much more compacted, although centripetal deposition around the channels to form primary osteons had not yet begun.

From the anatomical descriptions by Andrzej Elzanowski (1981) of *Gobipteryx,* the embryonic Cretaceous bird from Tsmeen Tsav, Mongolia, it appears that at least two stages of embryogenesis are reflected. The two better-preserved specimens described show differences in the level of skeletal ossification, notably in the vertebrae. In addition, although the "articular ends" of the tibia and femur are both preserved, they appear to be less ossified in one of the skeletons (Fig. 6.7). Histological sections of a tibial diaphysis of *Gobipteryx* show a more developed fibrolamellar bone (see Fig. 6.7) than the rather cancellous bone described in *Hypacrosaurus* by Horner and Currie (1994). This suggests that *Gobipteryx* is at a more advanced stage of development (but how this corresponds to embryonic age is uncertain). It is also possible that perhaps the nonavian dinosaur *Hypacrosaurus* and the avian *Gobipteryx* had completely different growth strategies.

Dinosaur eggs from the Early Jurassic of South Africa were described by James Kitching (1979) several years ago. Surface preparation revealed embryonic remains inside the egg structures, one of which suggested an embryo with a diapsid skull. The eggs are considered to belong to *Massospondylus*, a commonly occurring prosauropod dinosaur in the South African Red Beds (e.g., Kitching 1979). In 1999, James Kitching and I (unpublished manuscript) decided to revisit these eggs using CT scanning technology. Much to our disappointment, even though small bits of bones were visible on the surface of the rock, they could not be distinguished from the surrounding matrix in the scans. Our failure to get any images of the bones seems to suggest that the bones were not well ossified (especially because well-ossified adult material from the same horizons could be discerned even when they are not visible on the surface). If this is indeed the case, then these so-called prosauropod embryos are at a very early stage of embryogenesis.

The Auca Mahuevo embryos described by Luis Chiappe and his team (1998) are especially unique because, besides the skeletal remains, they also preserve skin impressions of the embryonic dinosaur (Fig. 5.1). It is interesting that the pattern of scales preserved in the embryos is similar to the pattern of scales on *Saltasaurus*, an Argentinian titanosaur. In my analyses of thin sections of the embryonic skin of the Auca Mahuevo sauropods, I was unable to locate any osteoderms (i.e., bone) within the skin. This suggests that, perhaps, as in many extant reptiles such as in crocodiles and the Gila monster, osteoderms develop posthatching in a one-to-one replication of the embryonic scale pattern. It is noteworthy that the embryonic skin did not show the raised central crests of the larger scutes of *Saltasaurus.* However, young hatchlings and juveniles of *Crocodylus niloticus* (the Nile crocodile) have relatively smooth textured osteoderms and scales, which become increasingly rugose with age, and also with increasing age, the larger dorsal scutes tend to develop a central crestlike morphology.

As with any egg-laying vertebrate, hatching marks the completion of embryonic growth. It is interesting to note that the fossil record has revealed both hatched and unhatched eggs of dinosaurs. Unhatched eggs are fairly well known, and here there is no doubt that the dinosaur encased by the unbroken eggshell is embryonic. There are however several finds of hatched eggs without embryonic remains. There are also tiny dinosaurs that are identified as embryonic because of their association with eggshell fragments, their relatively small sizes, and their distinctive textured bone surface (e.g., *Protoceratops* and *Dryosaurus*).

Posthatching Behavior

In the 1970s, Jack Horner and Bob Makela (1979) caught the world's attention with the description of the communal ground nesting sites of *Maiasaura,* a medium-sized, plant-eating dinosaur. Because juveniles and adults were found close to the nests, they suggested that *Maiasaura* parents cared for their young and fed and protected them. In another study, Jack Horner and Johns Hopkins University paleontologist David Weishampel (1988) examined the articular ends of the *Maiasaura* juvenile bones, and because they appeared relatively unfinished, they proposed that these baby dinosaurs were altricial. Broken eggshells in the nests and wear facets on the juveniles' teeth sug-

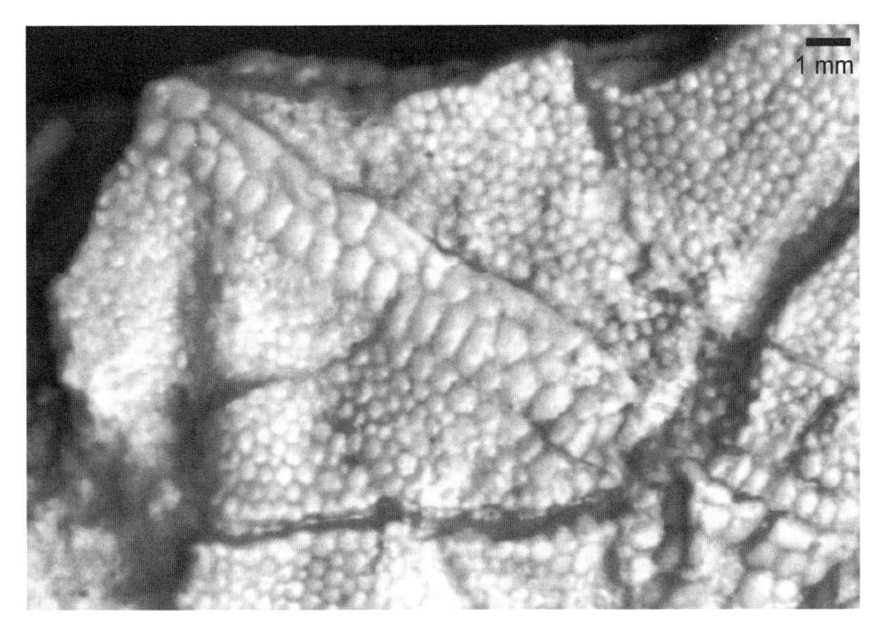

Fig. 5.1. Skin impressions of a sauropod embryo from Auca Mahuevo, an Upper Cretaceous locality in Argentina. Image courtesy of Luis Chiappe.

gested that the juveniles stayed in the nest and that food was brought to them. These finds were contrasted with the relatively unfragmented eggshells of *Orodromeus* (now identified as those of *Troodon*), the juveniles of which are considered to have been precocial. The question of precociality or altriciality of hatchling *Maiasaura* has been the point of much controversy. Wear facets on teeth of embryos are now known to develop even within the egg in crocodiles and it cannot be used to suggest that parents fed the young hatchling *Maiasaura*. Another problem with the altricial hypothesis is that as soon as the eggs of altricial birds hatch, the shells are either eaten by the parent or simply thrown out of the nest. However, they are not kept with the hatchlings in the nest, as is the case with *Maiasaura* nests.

The nature of the articular ends of dinosaur bone has been involved in the controversy of assessing whether dinosaurs were precocial or altricial. Jack Horner and David Weishampel (1988) examined the articular ends of baby *Maiasaura* and *Orodromeus* and deduced that the former had better formed articular ends than the latter. They proposed that the *Maiasaura* hatchlings were altricial, while the *Orodromeus* hatchlings were precocial. This is in direct contrast to the findings of Barreto and colleagues (1993) who showed that the articular surface of *Maiasaura* was more like that of precocial two-week-old chicks. It is worth noting that the articular ends of *Maiasaura* are not actually true articular ends because in the developing animal, the ends would have been mainly composed of unpreserved cartilage (only the calcified part of the end of the bone is preserved in fossil bones). Thus, surface texture of dinosaur hatchlings cannot be reliably used as an indicator of altricial or precocial behavior. More recently, Nick Geist and Terry Jones (1996), both former students of John Ruben in Oregon, have proposed that the degree of ossification of the pelvic girdle is a better indicator of altriciality or precociality. They have found that the pelvic girdle of perinatal *Maiasaura* is as well developed as that of postnatal crocodilians and precocial birds, which they suggest indicates that the *Maiasaura* hatchlings were precocial.

Discerning Hatchling and Juveniles

One of the important first-level deductions that needs to be made when a small dinosaur is discovered is whether one is dealing with a small-bodied taxon or a juvenile of a larger species. Hatchling and very young dinosaurs are in general easily identifiable because of the extremely fragile, spongy appearance of their skeletons. Large orbits, short snouts, large brain cases, and the lack of fusion of various skeletal elements typify the anatomical characteristics of juveniles. The latter are often reflected in the juvenile skeleton as unfused sutures on the skull, lack of fusion of neural arches to the vertebral centra, or the absence of fusion of scapulae to coracoids, as well as separate pelvic elements.

Fascinating studies examining the timing of skeletal fusion during ontogeny (i.e., when different parts of the skeleton fuse) have been undertaken by a number of researchers on a variety of living vertebrates. In a detailed analysis of crocodilian growth and assessment of how bones of the skeleton fuse during development, Chris Brochu (1996) found that the closure of the neurocentral sutures follows a caudal (tail) to cranial (head) sequence (i.e., the neurocentral sutures start closing from the tail region to the head region). He was able to pinpoint that sexual maturity in crocodilians occurred before the closure of sutures on the sacral vertebrae. In birds, it is recognized that the fusion of the proximal ends of the tibiotarsus and tarsometatarsus is a good indicator of adult status. Such clear-cut anatomical distinctions pertaining to the onset of sexual maturity are not known for dinosaurs. However, certain characteristics such as horns and frills (in the ceratopsians such as *Protoceratops* and *Chasmosaurus*) and cranial crests (in *Lambeosaurus*) may well be secondary sexual characteristics. It is possible that, given large enough sample sizes and the application of morphometric techniques and skeletochronology, distinctions of the actual ontogenetic timing of the development of such characteristics will be attainable.

Macroscopic observation of bone surface textures has been widely used to discriminate between different growth stages among extinct animals (e.g., ichthyosaurs, pterosaurs, pelycosaurs, and dinosaurs). Scott Sampson and colleagues (1997) used surface texture or "periosteal aging" of ceratopsian cranial elements to distinguish different ontogenetic stages. For example, juvenile skulls are distinguished by parallel striations oriented in the direction of growth, while a mixture of striated, mottled, and nonstriated bone textures typify subadults. Adults, however, are distinctive in having no striations, and the bone tends to vary from smooth to rough. Allison Turmakin-Deratzian of the University of Pennsylvania (2003) has recently completed a fascinating study of surface textural changes through ontogeny in crocodilians, birds, and dinosaurs. This research will no doubt allow an assessment of the usefulness of bone texture in determining ontogenetic status for dinosaurs, as well as for their closest living relatives.

Several growth series of dinosaurs are known, although some have better representation of different growth stages than others. For example, *Hypacrosaurus* is possibly the most complete growth series known (i.e., has good representation of all stages of growth). Several other dinosaurs have only partial growth series (e.g., *Allosaurus, Dryosaurus, Centrosaurus, Chasmosaurus, Coelophysis, Massospondylus, Protoceratops, Psittacosaurus,* and *Syntarsus*). Of these, some lack hatchlings and very young individuals (e.g., *Syntarsus, Massospondylus, Psittacosaurus*), while others are lacking in good representation of adult specimens (e.g., *Maiasaura*).

Notwithstanding the completeness, these growth series of dinosaurs have led to a considerable understanding of how dinosaurs grew. For example, trends during development, such as an increase in tooth counts (e.g., in nodosaurs, hadrosaurs, *Chasmosaurus, Psittacosaurus*), an increase in skeletal robusticity (e.g., *Syntarsus, Stegosaurus*), and a decrease in relative size of orbits (e.g., *Coelophysis, Psittacosaurus, Dryosaurus*), as well as the growth of crests in the lambeosaurines, are well recognized.

Bone Microstructure and Growth

Once the potential of using bone microstructure to interpret various aspects of the biology of dinosaurs was realized, the quest to assess dinosaurs' biology by analyzing their bone microstructure began in earnest. In the early part of the twentieth century, researchers were simply trying to document the nature of dinosaur bone tissues. Thus, our knowledge of the variety and type of bone microstructure among the Dinosauria continued to grow rapidly, even though most of these studies were mainly conducted on isolated fragments of dinosaur bone. Later, the study evolved from identifying well-preserved bone tissue in individual isolated fragments of bone to the histological changes through ontogeny within single species and in some cases within particular elements of single species, to multiple elements of single species.

As briefly mentioned in chapter 4, the earliest study using bone histology to distinguish ontogenetic status of dinosaurs was the 1933 study of Canadian hadrosaurs of the subfamily Lambeosaurinae by Baron Francis Nopsca (a rather eccentric Transylvanian paleontologist) and his colleague Heidsieck. The story unfolds in 1924 when the hadrosaur *Cheneosaurus* was then considered to be either a juvenile *Corythosaurus* or *Parasaurolophus.* A few years later, two skulls that were very similar to *Cheneosaurus* were ascribed to the new genus *Tetragonosaurus.* At the time, Baron Nopsca was working on reptilian bone histology, and he decided to undertake studies of the bone microstructure of fossil bone to assess whether, in fact, *Tetragonosaurus* was an immature hadrosaur or whether it was an adult and therefore a valid new genus. After procuring a piece of a rib of *Tetragonosaurus* and promptly making thin sections of the bone, Nopsca compared the bone microstructure of this dinosaur with that of other hadrosaurs, as well as to that of a variety of other dinosaurs, mammal-like reptiles, and ichthyosaurs. He found that the bone microstructure of the *Tetragonosaurus* rib was similar to that of a semiadult *Struthiomimus* and other small individuals of *Dicynodon, Lystrosaurus,* and *Ichthyosaurus.* He therefore concluded that the bone microstructure of *Tetragonosaurus* indicated a juvenile ontogenetic status for the dinosaur and that it, like *Cheneosaurus,* was an immature individual of some type of crested hadrosaur. (In 1975, Peter Dodson studied *Procheneosaurus* and found it and *Tetragonosaurus* to be juveniles of *Corythosaurus.*) For many years, the Nopsca and Heidsieck (1933) study remained the only ontogenetic study of dinosaurian bone microstructure.

In the late 1980s, Jill Peterson, a student working with Jack Horner in Montana, began working on the bone histology of a growth series of *Maiasaura peeblesorum.* Unfortunately, except for an abstract, she did not publish any of this work. However, almost 20 years since the onset of the *Maiasaura* bone microstructure studies, Jack Horner, Kevin Padian, and Armand de Ricqlès (2000) published a detailed assessment of the bone microstructure of *Maiasaura* through ontogeny.

In 1986, I began working on the bone histology of the small ceratosaur *Syntarsus rhodesiensis* (Fig. 5.2). Although *Syntarsus* is known from South Africa, I worked on the remains recovered from the Early Jurassic Forest Sandstone Formation of Zimbabwe. All the *Syntarsus* bones I used were recovered from a monospecific assemblage, which consisted of the remains of more than thirty individuals of different growth stages. The taphonomic indications as deduced by Mike Raath (1969) are that a catastrophe befell the social group and caused their mass death. While studying the skeletal remains, Raath (1990) became aware of characteristics such as trochanters and muscle scars in the femora that suggested sexual dimorphism in *Syntarsus.* He found that the dimorphism manifested itself as robust and gracile forms, with robusticity acquired only after sexual maturity. Raath proposed that the larger, more robust morph represented females, while the more gracile types were males.

Fig. 5.2. Gerard Marx's artistic depiction of the southern African dinosaurs, *Syntarsus rhodesiensis* and *Massospondylus carinatus.* Image courtesy of Gerard Marx.

When I began my studies of *Syntarsus* bone microstructure, I was well aware from the published literature that bone histology can vary significantly between different bones in the skeleton and within single bones as well. As such, to document histological changes through ontogeny in a particular species, one would need to section entire skeletons of different growth stages (i.e., from embryos to mature adults—a feat that no curator would allow especially since, from firsthand experience, I know how difficult it is to obtain permission to section even a single element comprehensively!). The next best thing was therefore to standardize the study by examining homologous elements of different growth stages of *Syntarsus.* Since long bones are preferred over the highly remodeled ribs, pelves, scapula (shoulder blades), jaws, and most other skeletal elements, I focused on the femora of *Syntarsus.* To constrain the variables even further, I sectioned the so-called neutral zone in the midshaft region of the femora, which is the area least affected by remodeling changes and therefore most likely to retain most of the bone microstructure through ontogeny. (See chapter 2 for more information concerning sample selection for skeletochronology.)

In a study of *Syntarsus,* I examined twelve femora ranging from 12 to 23 cm in length (which is the largest-sized femur known). Like the bones of modern mammals and birds, I found that *Syntarsus* bone was richly vascularized and consisted of a large amount of fibrolamellar bone. However, like crocodilians, this period of fast growth was periodic and alternated with distinct periods of slowed growth and periods of arrested

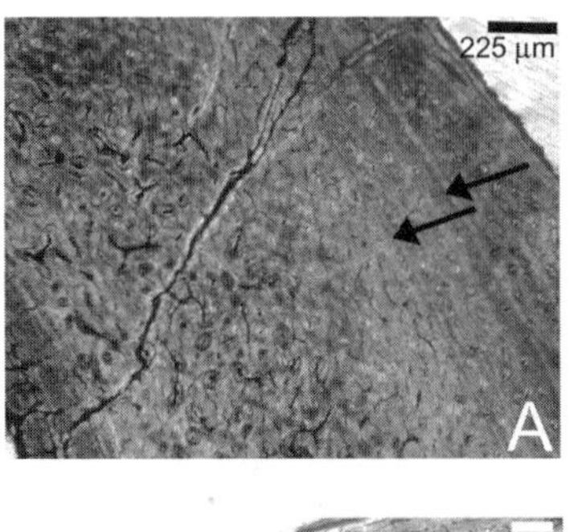

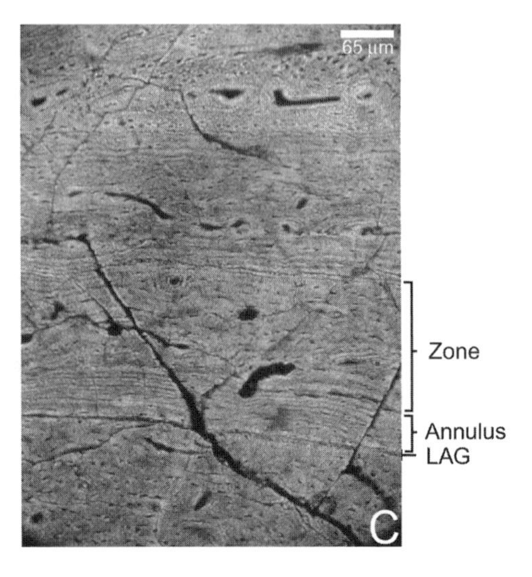

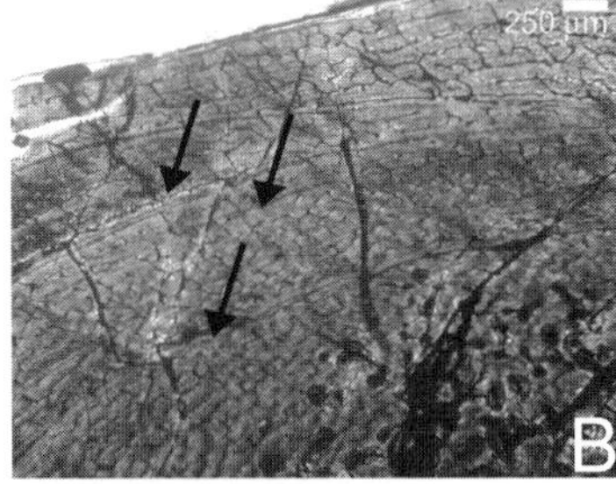

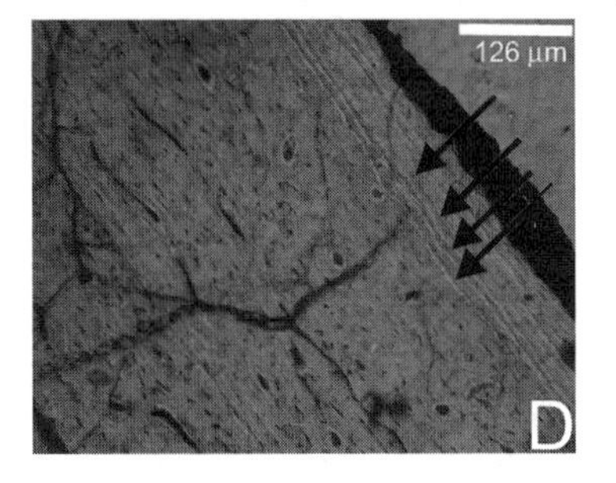

Fig. 5.3. *Syntarsus* bone microstructure. *A* and *B* show the zonal nature of the compact bone. Arrows indicate the growth rings. *C*. Higher magnification showing microstructural details of growth rings. *D*. Rest lines in the peripheral region of an adult bone. Compare with rest lines in a dog illustrated in Fig. 3.9*D*.

growth (Fig. 5.3). Thus, the whole compact bone appeared to have had a stratified nature because of the zonal bone. Furthermore, I found that with increasing size the number of growth rings increased. By applying skeletochronology to the growth marks in the bone, I was able to determine a growth trajectory for *Syntarsus* (Fig. 5.4).

This was a significant contribution to dinosaur paleobiology because until this study, longevity and growth rates of dinosaurs were much debated and based only on indirect and sometimes theoretical models or frank "guesstimates." Depending on researchers' perceptions of whether dinosaurs were endotherms (warm-blooded) or ectotherms (cold-blooded), several decades to a single year have been proposed as the time required for large dinosaurs to reach maturity: Ted Case (1978) of the University of California, San Diego, proposed that large dinosaurs required about 60 years to reach maturity. Independent researcher Greg Paul (1988) of Maryland has proposed that dinosaurs grew rapidly like mammals and birds. Between these extremes, Arthur Dunham of the University of

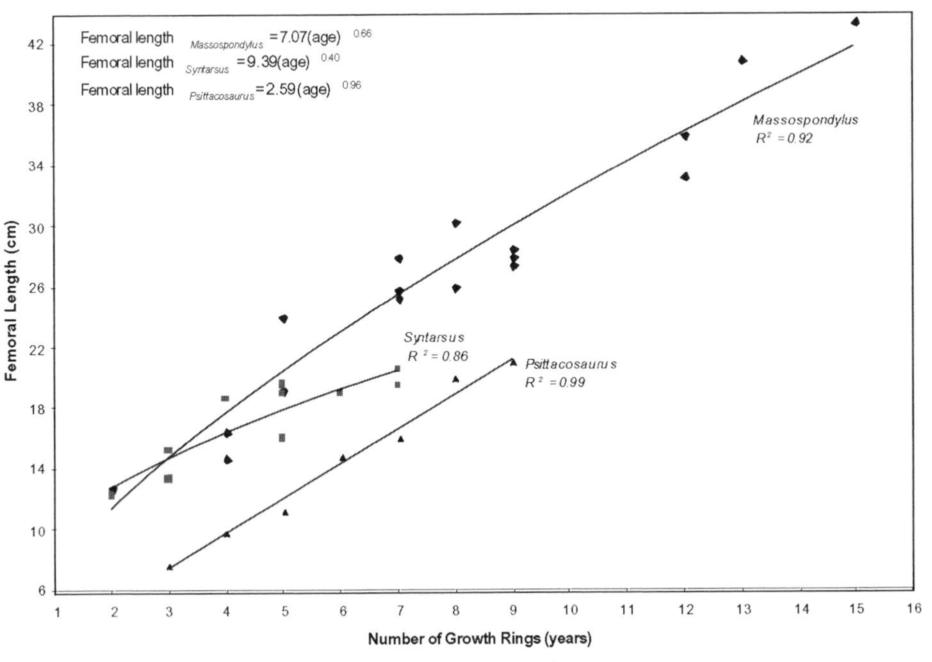

Fig. 5.4. Growth curves of *Syntarsus, Massospondylus,* and *Psittacosaurus* based on femoral growth. Data for *Syntarsus,* from Chinsamy (1990), for *Massospondylus* from Chinsamy (1991, 1993a), and for *Psittacosaurus* from Erickson and Tumanova (1999).

Pennsylvania (1989) and his colleagues, using theoretical modeling, suggested that the slow growth rates extrapolated from living reptiles for dinosaurs would be highly improbable because this would require extremely high juvenile survivorships. Instead, considering the nature and magnitude of various demographic and physiological factors that constrain life history variation, a more reasonable estimate for growth to adult size for a large hadrosaur, and for other large dinosaurs, would be about 10–12 years. My findings using skeletochronology suggest that *Syntarsus,* weighing probably about 20–25 kg, took about seven to eight years to reach a mature body size. In *Syntarsus,* ontogenetic maturity was established because of the closely spaced rest lines in the peripheral part of the compact bone. As mentioned previously, in mammals and birds, this feature suggests that growth has slowed down and that a mature body size had been attained. Thus, I am reasonably sure that in the largest individual of *Syntarsus* the slow accretion of bone suggests maturity and hence a determinate growth strategy for the taxon. Although Mike Raath (1990) had distinguished robust and gracile morphs (possibly representing male and female) of *Syntarsus,* separate growth curves for each of the morphs were not possible because the robust morph is only distinctive at sexual maturity (i.e., juveniles are indistinguishable), and only one gracile adult morph was available for histological analysis.

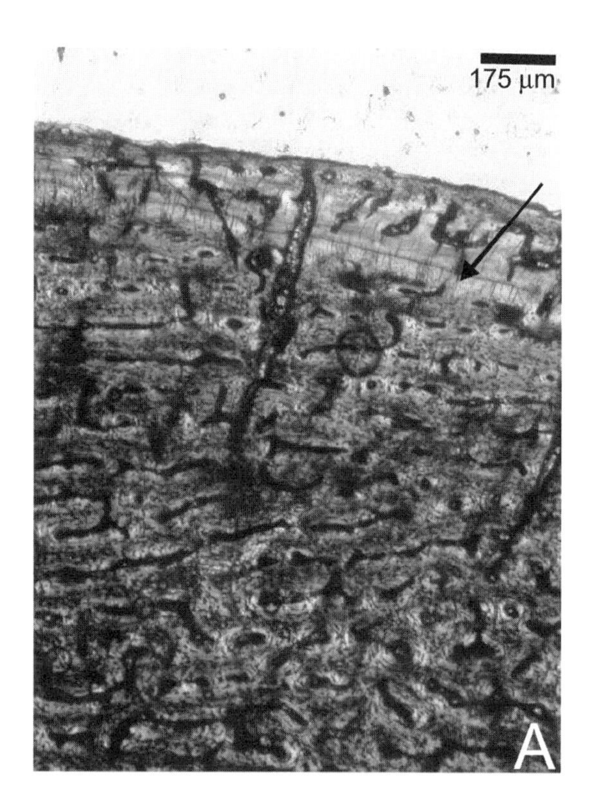

Fig. 5.5. *Massospondylus* bone microstructure. *A*. Juvenile femur measuring 12.75 cm showing one growth ring. *B*. Adult bone showing several growth rings. Arrows indicate the growth rings in the bone.

I also worked on the growth series of *Massospondylus carinatus,* a prosauropod dinosaur, also from the Early Jurassic of South Africa. Here, the sample consisted of nineteen femora with a femoral length size range of about 12–44 cm. Histological examinations showed that the bone was deposited in a cyclical manner with definite periods of fast growth indicated by the fibrolamellar bone alternating with lines of arrested growth (LAGs) and annuli, which consisted of a lamellar type of bone tissue (Fig. 5.5). As with the contemporaneous *Syntarsus,* these specimens also showed an increase in the number of growth rings with increase in size. The amount of bone deposited between consecutive rings was quite variable across a cross section and depended on the local morphology of the bone (Chinsamy 1993a). The largest femur (length of 44 cm) is estimated to have had at least fifteen growth rings. Counting the number of growth rings and plotting them against the femoral length led to the deduction of growth curves for the dinosaur (Fig. 5.4). Note, that in some instances the same number of growth rings is evident in the compacta of different-sized femora. For example, seven growth rings occur in femora that are 27.92 cm, 25.23 cm, and 25.84 cm long, while eight growth rings were documented for femora that were 25.95 cm and 30.3 cm in length. This variation could be the result of gender differences in the sample (which are not discernible morphologically) or perhaps simply a result of individual variation. Nevertheless, the growth curve depicted averages in the variation in the data and results in an approximation of the growth strategy for the taxon. The projected growth trajectory suggests a growth rate intermediate between that of reptiles and mammals and birds.

In an independent study, David Varricchio (1997), working in the Museum of the Rockies in Montana, examined the ontogenetic growth of *Troodon formosus* by analyzing the bone microstructure of three third-metatarsals and two tibiae of different sizes. Considering the number of lines of arrested growth in the compacta of the large individuals studied, Varricchio found that this small Upper Cretaceous theropod, weighing about 50 kg, took between 3 and 5 years to reach adult size. Varricchio also suggested that bone formation in *Troodon* went through three stages, that is, a rapid fibrolamellar, moderate lamellar-zonal, and a slow avascular lamellated stage. The last stage suggests a determinate growth strategy. It is regrettable that a series of femora was not available, making precise comparison with earlier studies difficult.

During 1992, while undertaking postdoctoral research with Peter Dodson at the University of Pennsylvania, we managed to obtain a growth series of *Dryosaurus lettowvorbecki,* a medium-sized ornithopod dinosaur from the Late Jurassic deposit of the Tendaguru Beds of Tanzania from Frank Westphal of the University of Tübingen (Chinsamy 1995b). The bone microstructure of *Dryosaurus* was different from any of the other growth series previously studied. Unlike *Syntarsus, Massospondylus,* and

Troodon, Dryosaurus appeared to grow at a sustained rapid rate, without any interruptions in the rate at which it formed bone (Fig. 5.6). In all thirteen femora of different sizes examined, no annuli or lines of arrested growth were present. Although no growth marks were formed, it was clearly evident from the nature of the bone tissue that during early ontogeny, bone was deposited at a faster rate than during later ontogeny. In the publication of *Dryosaurus* bone histology in 1995, I deduced the minimum age of the youngest individual as 80 days, by using bone depositional rates and rates of secondary osteons formation of mammals. I now realize that the rate at which lamellar bone forms can be quite variable in different animals, and that though this was a novel idea, it failed to consider the nuances of bone depositional rates.

Another interesting aspect of *Dryosaurus* bone microstructure is the fact that some distal and proximal cross sections of the femora consisted almost entirely of compacted coarse cancellous bone. As mentioned in chapter 3, this condition was not previously reported for dinosaurs, although it is known from mammalian bones and suggests that these regions were in the medullary region (as cancellous bone) at an earlier stage of development and that during growth of the femur, it became relocated into the shaft of the bone (see Fig. 3.1).

Recent work on a growth series of *Coelophysis* femora shows that the compacta is dominated by a richly vascularized fibrolamellar bone tissue. However, in several sections, this rapidly formed bone tissue is distinctly interrupted by growth rings. *Coelophysis* and *Syntarsus* are considered as closely related and Greg Paul (1998) has suggested that they belong to the same genus. A preliminary comparison of their bone structure shows that microstructure of the compacta of *Coelophysis* is different from that of *Syntarsus* in that it has less intensive secondary reconstruction and fewer secondary osteons. Furthermore, in *Coelophysis,* the LAGs are not followed or preceded by an annulus, which suggests that the interruptions in bone formation and hence growth were fairly abrupt. Preliminary findings suggest that the relative bone wall thicknesses of comparably sized *Coelophysis* specimens are much thinner than that of *Syntarsus.* This is an interesting finding, given that exercised piglets, that is, those that ran on a treadmill for 30 min each morning and afternoon at about 3 mph, have thicker cortical bones than siblings that were kept in pens where they could only walk about (Lieberman and Pearson 2001). One could speculate that *Syntarsus* experienced more strain in its limb elements than *Coelophysis*, but this work is still very much at an early stage and it is too premature to draw any deductions.

In 2000, Greg Erickson and Russian paleontologist Tatyana Tumanova from the Paleontological Institute of the Russian Academy of Sciences published an account of the bone histology and growth of the Late Cretaceous ceratopsian, *Psittacosaurus mongoliensis.* Seventeen skeletal elements comprising humeri, femora, and tibiae at varying

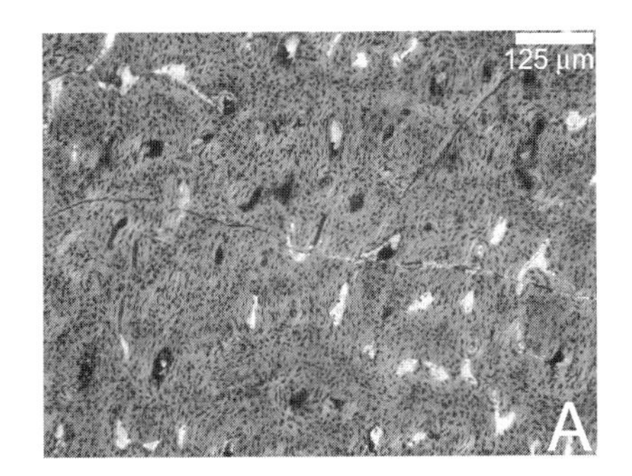

Fig. 5.6. Bone microstructure of *Dryosaurus* from Tanzania. *A*. Fibrolamellar bone in a juvenile. *B*. Azonal bone in a subadult individual. *C*. Azonal bone and no rest lines in the peripheral region of an adult femur.

stages of ontogeny were studied. They found that the bones consisted predominantly of fibrolamellar bone that was interrupted periodically by annuli consisting of parallel-fibered bone and/or a lamellar bone matrix. Seven femora of different sizes (ranging from 7.6 to 21 cm) were included in the study, and each femur showed a distinctive number of growth rings (unlike the case in *Massospondylus* and *Syntarsus* [see Fig. 5.4], where different-sized bones had similar numbers of growth rings). On the basis of growth ring counts, the largest *Psittacosaurus* femur was estimated as 9 years old, and *Psittacosaurus* is considered to have taken 9 years to reach a body size of 19.87 kg. When femoral length of *Psittacosaurus* is plotted against age (using data from Erickson and Tumanova 2000), it suggests a constant growth rate throughout ontogeny, without any sign of plateauing (Fig. 5.4).

In the *Psittacosaurus* study, the researchers developed a novel methodology called "developmental mass extrapolation" (DME) to deal with the problem of deducing the mass of dinosaurs through various stages of ontogeny. Growth rates per day were determined for *Psittacosaurus* and plotted on a graph showing growth rates of extant vertebrates. Erickson and Tumanova (2000) deduced that *Psittacosaurus* maximal growth rates exceeded extant reptiles and marsupials but were slower than modern birds and eutherian mammals. In fact, looking at the data, they could quite accurately have said that *Psittacosaurus* growth rate fell quite close to that of marsupials (Fig. 5.6A).

In recent years, there have been a few fascinating ontogenetic studies on sauropod bone microstructure and life history: Frédérique Rimblot-Baly and colleagues (1995) on *Lapparentosaurus,* Kristina (Kristi) Currey-Rogers (1999) on *Apatosaurus,* and Martin Sander (2000) and Sander and Tückmantel (2003) on the Tendaguru sauropods.

The Rimblot-Baly study expanded on the early work by de Ricqlès (1983) on *Lapparentosaurus* (=*Bothriospondylus*), a Madagascan Middle Jurassic sauropod and involved an analysis of ten humeri that ranged in size from 30 to 155 cm in length. The researchers documented that bone deposition was high in early ontogeny, but that later in ontogeny, the dinosaur grew at a slower rate and in an apparent cyclical manner. Given that the epiphyseal sections showed active longitudinal growth, they deduced that the sauropod probably experienced indeterminate growth. Furthermore, histological findings suggest that *Lapparentosaurus* took about 20 years to reach adult status, although longevity could not be assessed. Using the thickness of the cortical bone, divided by the estimated number of growth rings, the researchers suggest that the laminar fibrolamellar bone formed at about 7 microns per day in *Lapparentosaurus* (Rimblot-Baly et al. 1995).

In 1999, Currey-Rogers, then at the University of Stony Brook, New York, studied the bone microstructure of five ulnae, four radii, and three scapulae of *Apatosaurus.* The skeletal elements were from four individuals at different ontogenetic stages (though, except for one adult in the sample, and a scapula that is 34% of the adult size, the data set appears to be skewed toward the subadult size range, i.e., between 56% and 91% of adult size). Currey found that below about 75% of adult size, fibrolamellar bone was the dominant tissue type in all the skeletal elements. However, significant differences were reported between the forelimb and girdle elements: the radii and ulnae formed laminar bone at sustained rates, while at low magnification, the scapulae appeared to show stratification on the basis of variation in vascular orientation and density (i.e., regions of relatively slower growth exhibit longitudinal vascular canals, while regions of faster growth show more extensive vascular networks). It is interesting that no annuli or LAGs are observed, which suggest that such variation represents mild modulations in the rate of bone formation. After attaining about 75% of adult size, the rate of bone formation slows down in all skeletal elements of *Apatosaurus* studied, and annuli, LAGs, and parallel and/or lamellar-fibered matrices are observed in the bones. This led Currey to suggest that *Apatosaurus* had determinate growth. However, slowed growth is not exactly the same as determinate growth. This is especially true for crocodilians that grow slowly after sexual maturity but still increase their lengths by more than 50%. It is interesting that Martin Sander (2000) has found prolonged slow growth for *Janenschia*. It should also be noted that cross-sectional appositional growth does not reflect growth in length, and as in other studies, no firm deduction can be made regarding determinate or indeterminate growth until distal and proximal ends of adult bones are studied. However, on the basis of cross-sectional histology, it is reasonable to deduce that appositional diaphyseal growth had slowed down in *Apatosaurus* after about 75% of adult size.

On the basis of the vascular cycles in one scapula (56% adult size), Currey (1999) deduced that a growth plateau followed a period of rapid growth at around 8–10 years when subadult size was reached. I find this deduction problematic. First, the 34% adult-size scapula shows five vascular cycles, while the 56% adult-size scapula shows eight to ten cycles, and the adult-size scapulae shows none of the fibrolamellar bone of the earlier stages, and here, the periosteal bone shows distinct zones with LAGs. However, when deducing age to maturity, Currey (1999) only uses the eight to ten cycles in the second age class scapula, assuming that no cycles were obliterated between the two growth stages. Just looking at the substantial increase in size of the scapulae and the medullary cavity, it is obvious that substantial remodeling must have occurred. I believe that the deduction of 8–10 years to maturity is underestimated. Furthermore, from the description of the adult scapula that Currey provides in her publication on *Apatosaurus* bone histology, it appears that several more zones and LAGs were deposited during scapular growth to adult size. Because scapulae have a complex morphology, they require substantial remodeling during growth, and as a consequence, much of the history of bone growth is obliterated, which I believe makes them unreliable for assessing growth dynamics of dinosaurs. As a result, I am skeptical about the data obtained by Currey (from just one bone) and feel that not enough attention has been given to the amount of reconstruction that has occurred. Interestingly, in a similar-sized *Apatosaurus* (to the one that Kristi Currey-Rogers worked on), Martin Sander and Christian Tückmantel (2003) found twenty-five polish lines in a femur and also estimated (using bone depositional rates) a minimum age of 33 years, which concurs with my deductions that Currey's (1999) age assessment is underestimated.

It is interesting that Haversian bone occurred in all the *Apatosaurus* bones studied and is especially abundant in older individuals. The lack of individuals with just primary compact bone either suggests that the early juveniles are not that young or that for the diplodocids, secondary reconstruction begins fairly early in ontogeny.

In 2000, Martin Sander, from Bonn University in Germany, published an account of the bone microstructure of four Upper Jurassic sauropods from Tendaguru in Tanzania, Africa. He examined a growth series of about fifteen femora and humeri of *Brachiosaurus* and *Barosaurus,* while for *Dicraeosaurus,* he had a sample of six individuals with a more limited size variation, and for *Janenschia* only a humerus and femur were available for study. Instead of attempting to cross section the large sauropod long bones, and because of the destructive nature of such a methodology, Sander took bone core samples from the diaphyseal regions of the long bones, with a five-eighths-inch-thick drilling bit. (Although practical, this methodology presumes that the bone microstructure is uniform in the cross section of the bone—in reality, this is rarely the case.) Sander was able to recognize two types of growth marks in sauropod bones: the usual LAGs that mark periods of arrested growth and another growth mark that he describes as "polish lines," which he defines as growth lines in fibrolamellar bone that are visible in polished sections with the naked eye (but are not recognizable in thin section). Sander suggests that the polish lines may represent a slow down in growth, which is not pronounced enough to cause a LAG to be formed.

One of the most remarkable finds of Sander's (2000) study is the distinctive histology associated with each of the four sauropod taxa. In the *Barosaurus* sample, Sander found that five humeri, four femora, and a fibula showed a bone microstructure that indicated rapid uninterrupted growth with a very limited amount of remodeling, while a humerus, three femora, and a tibia showed a slower, interrupted pattern of bone formation that was accompanied by extensive remodeling. Sander recognized that such differences may be a result of a mixture of two taxa in the sample but proposes that the second type of bone tissue could represent females of the species and that the discontinuities in growth could be linked to reproductive demands on the female skeleton. As discussed in the chapter 4, I think that the sample size is much too small to make this sort of deduction, and it possible that two taxa rather than a male and female morph are represented in the sample. The *Brachiosaurus* bones studied showed a thick cortex of fibrolamellar bone, with substantial remodeling in older individuals, while *Dicraeosaurus* showed a high degree of secondary reconstruction. *Janenschia* was most different in that it exhibited a thick cortex with regularly spaced polish lines.

As Currey (1999) found in *Apatosaurus,* fibrolamellar bone tissue appeared to be the dominant tissue type also among the Tendaguru sauropods. In all the taxa examined, growth appeared to slow down drastically once individuals reached sexual maturity, and except for the "female morph" of *Barosaurus* and *Janenschia,* growth appeared to be rapid without periodic interruptions. Since *Janenschia* showed regularly interrupted cycles in growth, Sander estimated that the animal reached sexual maturity at about 11 years, with growth essentially reaching a plateau at about 26 years though the animal continued growing at a much slower rate until about 38 years. Thus, the plateau of growth at about 26 years is more than double that which was reported for *Apatosaurus* by Currey (1999).

In 2000, Jack Horner, with his collaborators, Armand de Ricqlès and Kevin Padian, published the long-awaited detailed account of ontogenetic growth in *Maiasaura.* In this study of six growth stages (early and late nestlings, early and late juveniles, subadult and adult) of a variety of skeletal elements, they were able to document the histological changes at each stage. On the basis of the type of tissue present at each stage, they deduced that young *Maiasaura* nestlings (about 45 cm in length) grew at very high rates and grew at high and moderately high rates during later nestling, juvenile, and subadult stages, with adults having the lowest growth rates. They further deduced that adult size of 7 m was attained in 6–8 years. The latter was deduced in two ways: by skeletochronology (i.e., a count of the number of LAGs in the single tibia that was representative of adult *Maiasaura)* and by using bone depositional rates obtained from modern animals. However, both deductions are problematic. Skeletochronology on a single element is unreliable, and bone depositional rates of modern animals cannot be extrapolated to dinosaurs because similar bone tissue types do not imply similar bone depositional rates (Starck and Chinsamy 2002; de Margerie et al. 2004). Until such time as more detailed analyses are performed on a wide variety of vertebrates, it is much too premature to extrapolate modern bird bone depositional data to dinosaurs, as in the case of the *Maiasaura* study, where the authors assumed consistent depositional rates, for example, 30 microns per day for radial growth between hatchling and large nestling or a constant rate of 15 microns per day for later stages. Given the variation in bone depositional rates with ontogenetic age, skeletal element, and environmental constraints from just the handful of studies on extant vertebrates, it is inadvisable to apply such data to dinosaurian bone growth.

In 2001, Erickson, Currey, and Yerby provided some preliminary data on three femora of *Shuvuuia,* a derived maniraptoran theropod. Details of this study are not yet available, but from the single image provided in the publication, it looks as though

Shuvuuia has a fibrolamellar type of bone tissue, with distinct periodic interruptions in the rate of bone formation.

Recently, two studies of the growth dynamics of *Tyrannosaurus rex* were conducted (Erickson et al. 2004; Horner and Padian 2004). Applying skeletochronology on non-weight-bearing bones (see chapter 4), Erickson and colleagues (2004) deduced that *T. rex* reached skeletal maturity by 18.5 years and probably lived to about 28 years. Horner and Padian (2004) reached a similar conclusion based on a single fibula of an adult in which only half the compacta preserves primary bone tissue with 17 LAGs. The rest of the growth lines were back-calculated to suggest that the individual was between 22 and 25 years old (although no consideration was made of bone removed during earlier growth stages). Erickson and colleagues (2004) also deduced that the smaller bodied tyrannosaurid taxa, *Gorgosaurus liberatus*, *Albertosaurus sarcophagus*, and *Daspletosaurus torosus* attained somatic maturity in about fourteen to sixteen years.

At present, no growth series of dinosaurs has been studied from embryonic stages through to mature adult stage, primarily because very few taxa have such well-represented growth stages. Of the few studies of ontogenetic changes in bone microstructure, only that of *Maiasaura* includes hatchlings; however, only one adult tibia is represented in the sample. The other growth series studies on *Syntarsus, Massospondylus, Troodon, Dryosaurus,* and *Psittacosaurus* comprise mainly late juvenile, subadults, and adults stages.

One of the most complete dinosaur growth series is that of *Hypacrosaurus stebingeri* (Ornithischia, Lambeosaurinae) from the Uppermost Two Medicine Formation of northern Montana and southern Alberta. This material represents the largest collection of embryos and nestlings and includes juveniles and adults. Jack Horner and Phil Currie (1994) found that the embryonic compacta has a woven bone matrix with abundant osteocyte lacunae, although primary osteons are still in early stages of formation. Nestling samples show that the osteons are better developed and the fibrolamellar nature of the bone can be ascertained. Histological analyses of the ends of long bones show abundant calcified cartilage arranged in columns. These histological findings suggest that the embryos and nestlings were growing rapidly. Histological study of this exceptional growth series of *Hypacrosaurus* will offer the most complete account of dinosaur growth and it is anticipated that such a study will be forthcoming in the near future. Combined with the allometric and morphometric studies, *Hypacrosaurus* will no doubt add substantive information to our understanding of how dinosaurs grew.

Embryonic Bone Tissues

Until recently, except for *Hypacrosaurus,* our knowledge of dinosaur embryonic bone tissue was fairly limited. More recently, Horner and colleagues undertook an extensive study of embryonic and perinatal dinosaurs, living birds, and modern reptiles. They documented distinctive types of histology for nonavian dinosaurs and other reptiles. These researchers reported strong phylogenetic signals in the distribution of bone tissues and vascularity of long bone shafts and epiphysis. They found that a 55-day-old snapping turtle showed thin-walled limb bones, with a porosity (equated to "vascularization") of about 12% and no fibrolamellar bone present. A full-term embryonic alligator had similar tissues to the turtle, with a porosity of 14% and with some fibrolamellar bone deposited. However, porosity is reduced to 4% in alligators less than one year old. In my studies on crocodilians (Chinsamy 1993b), I found that porosity values were quite variable in the compacta from the peripheral region of the bone wall to the perimedullary regions—this variation was not considered by Horner and colleagues (2001). I found midcortical porosity values of 4.6% for juveniles, but this was reduced to 1.85% in an almost 8-year-old crocodile. The modern bird taxa studied by Horner and colleagues (2001) have relatively high vascularized bone walls (porosity of 29%–32%), with variable types of fibrolamellar bone tissue present. Embryonic nonavian dinosaurs had thick bone walls, with porosity values of between 20% and 36% and poorly developed marrow cavities.

Growing Tall

To assess how dinosaur femora grew in length, it is necessary to study the "articular ends" of the skeletal element. One should be careful not to confuse this terminal surface of the fossil bone with the actual articular surface in the living dinosaur. In the fossil, the thick pad of uncalcified articular cartilage and articular cartilage that would have overlain this surface is not preserved. Therefore, the ends of an immature fossil bone reflect the junction between calcified and uncalcified cartilage.

The amount of calcified cartilage in the ends of long bones of dinosaurs can provide a reasonable assessment of whether growth in length was rapidly underway. Robin Reid (1997b) has proposed that because the shape of the ends of young dinosaur long bones tends to look like what would have been the articular ends of the bone, this suggests that growth in length occurred under the articular cartilage (i.e., that dinosaurs had a subarticular style of endochondral growth, as in crocodilians). This would have allowed the animals to have indeterminate growth (i.e., they would continue growing throughout their lives, as originally proposed by Armand de Ricqlès. If this were true, we would always expect to see calcified cartilage in the ends of long bones of dinosaurs, irrespective of how big they were, perhaps with decreasing amounts of calcified cartilage. This has not been adequately tested; although, many dinosaurs show definite slowing down in appositional growth of the diaphysis, whether this corresponds to a decrease in the amount of calcified cartilage has not been examined.

However, my own work on growth series of *Syntarsus* and *Massospondylus* does suggest a decrease in the rate at which the femora grows in length during ontogeny.

In a recent study, Horner and colleagues (2001) undertook a detailed comparative analysis of epiphyseal growth of embryonic and perinatal dinosaurs (*Orodromeus, Maiasaura, Hypacrosaurus, Troodon,* and an unidentified lambeosaurine), modern birds, and reptiles (snapping turtle and alligator). They found that the limb bones of perinatal *Orodromeus* and embryonic *Troodon* show thick pads of calcified cartilage penetrated by marrow tubules, and no recognizable secondary center of ossification. The researchers have argued that the tubes that open to the "epiphyseal" surface and are infilled by sediment reflect "cartilage canals." In epiphysis of embryonic theropod bones from Portugal, de Ricqlès and co-workers have also reported similar epiphyseal bone tissues. Horner and colleagues (2001) have suggested that these "cartilage canals" are like those of modern birds and are associated with rapid growth. However, as Haines showed in 1942, cartilage canals are not unique to birds and to eutherian mammals but are also well developed in epiphyseal regions of *Varanus exanthematicus,* although in the former cases, they are not as extensively branched and also extend directly into the growth cartilage. Distinguishing between cartilage canals and marrow tubules is quite difficult (Horner et al. 2001), and in thin sections of juvenile dinosaur epiphyseal regions that I have studied, I cannot with confidence verify that cartilage canals occur. However, because Horner and colleagues (2001) have studied very young dinosaurs, perhaps in dinosaurs, the penetration of the growth plate by cartilage canals happens only at very early stages of growth and may aid initial rapid longitudinal endochondral growth. It is clear that more research is needed to understand whether in fact dinosaurs grew in length as crocodiles do (which Robin Reid has suggested) or whether in fact they grew as birds do (as Horner and colleagues propose).

Horner and colleagues (2001) also reported that in *Orodromeus,* endochondral bone extends right up to the proximal end of bone. The embryonic skeletal elements of *Troodon* and *Orodromeus* have substantially more endochondral bone formed than in the perinatal *Maiasaura*. In the latter, it is reported that the "marrow tubules" and "cartilage canals" meet to form long tubular structures along the walls of which bone is deposited. Whether such differences reflect precocial or altricial behavior or simply differences in the embryonic stage of development or differences in relative growth to adult size is still uncertain.

How Fast Did Dinosaurs Grow?

Studies of bone microstructure have provided valuable insight into how particular dinosaur taxa grew. The key question is, however, whether these deductions can now be extrapolated to the Dinosauria as a whole.

My work on growth series of three dinosaur taxa have led me to propose that although these dinosaurs did not grow as slowly as typical reptiles, their growth rates were not as great as those of mammals or extant birds. In several publications, I have made the case for an intermediate type of growth rate for dinosaurs and have proposed that given the diverse nature of the Dinosauria, it is quite likely that different taxa had different growth strategies (e.g., Chinsamy and Dodson 1995).

Several growth series of dinosaurs undeniably show that, although they form fibrolamellar bone abundantly in their skeletons, many also show periodic interruptions (pauses or slowing down in growth) in the form of annuli or LAGs. Horner and colleagues (1999) consider these LAGs, simply as a plesiomorphic characteristic, without any functional or physiological significance. These researchers suggest that only the predominance of fibrolamellar bone should be considered and suggest that dinosaurs grew at rapid rates comparable with modern mammals and birds. However, the proposition that plesiomorphic features are automatically without functional significance is clearly preposterous. Consequently, I believe that for any real discussion of the significance of dinosaurian bone tissues, the entire range of bone tissues, including features such as LAGs and annuli, must be considered.

It is interesting that not all dinosaurs show the same type of bone tissue. Much variation exists; for example, some taxa form bone at a fairly homogenous uninterrupted rate throughout ontogeny, whereas in others, bone formation slows down only later in life, and sometimes LAGs and annuli develop; still others form LAGs and annuli throughout their lives. Given such variation, it is obvious that there is no typical growth strategy representative of the Dinosauria. Matthias Starck and I (2002) have explained the variation observed in dinosaurs as a function of the plasticity of bone. It appears from the nature of bone microstructure of many extant and extinct archosaurs that the plesiomorphic condition is flexible growth in which rates of bone deposition (and growth) can fluctuate, while some members of the Dinosauria (e.g., some ornithopod dinosaurs) and the ornithurine birds have lost such flexibility.

Greg Erickson, Kristi Currey-Rogers, and Scott Yerby (2001) have also grappled with this problem of dinosaurian growth rates. They combined bone histology studies with "developmental mass extrapolation," which allows the calculation of body mass during ontogeny, so that growth curves (mass vs. age) could be plotted and compared with that of modern vertebrates (Fig. 5.7).

Unfortunately, there are several problems with their approach. A significant problem with DME-deduced growth rates is that a mixture of methods for estimating mass is used, which undermines the accuracy of results. For example, whereas in some cases (e.g., for *Apatosaurus*) age is estimated from scapulae (which, as I mentioned earlier, are

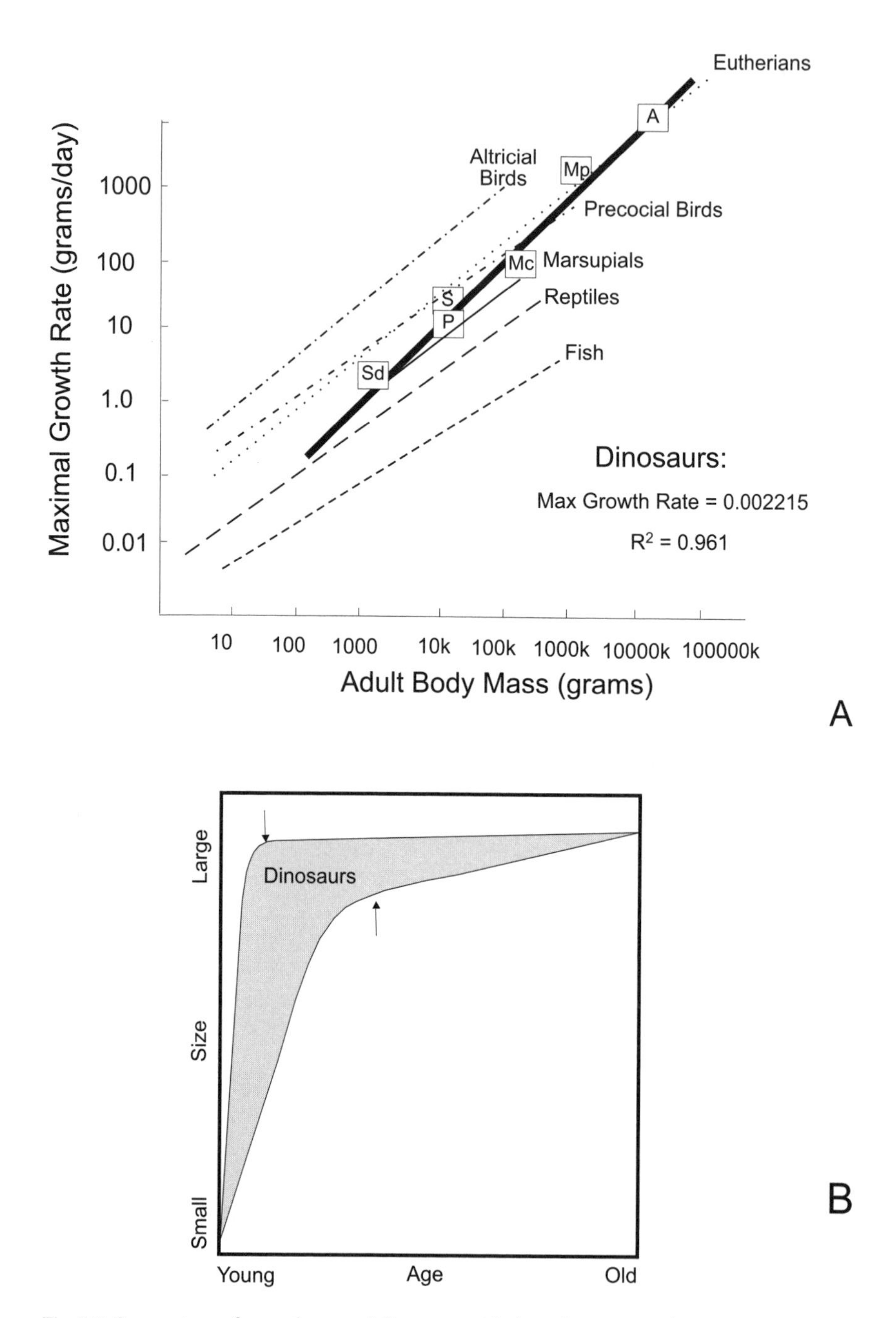

Fig. 5.7. Comparison of growth rates of dinosaurs with that of extant vertebrates (redrawn from Erikson et al. 2001). *A. Apatosaurus*; Mp, *Maiasaura peeblesorum*; Mc, *Massospondylus carinatus*; S, *Syntarsus rhodesiensis*; Sd, *Shuvuuia deserti*; P, *Psittacosaurus mongoliensis*. *B.* Growth rates of dinosaurs were probably intermediate between reptiles (bottom curve) and birds (redrawn from Chinsamy and Dodson 1995). Arrows indicate mature body size.

not reliable elements for skeletochronology), and the derived age therefore appears to be an underestimate, in other cases, long bones such as tibiae and femora are used. Moreover, for *Maiasaura* and *Shuvuuia* only three data points are used. Although the correlation coefficients presented are high, with so few data points, almost any curve would fit the data equally well.

Likewise, when the mass of animals is calculated, different formulae are used for different ontogenetic ages. For example, for adult mass, the Anderson and colleagues (1985) formula is used, which is derived largely from extant quadrupedal mammals. This presents a problem because mammals are not good models to assess dinosaur mass (especially for bipedal forms). In contrast, for younger animals (juveniles and subadults), mass estimates are derived from crocodilian models. For example, the femur length at each growth stage of *Psittacosaurus* is cubed because crocodilian allometric growth suggests that femoral length scales isometrically to about the 0.33 power during ontogeny. This mixture of methodologies conflates the final data and leads to a circular argument. Further compounding these problems, Erickson and colleagues (2001) only present a single growth rate estimate for each taxon without stating to which growth stage it applies. This is not reasonable because growth rates in general vary with age, and any one rate is only true for a particular stage of the animal's life.

Another problem with the DME method as applied by Erickson and colleagues (2001) is that it assumes that the largest femur is the adult size, and all the subsequent mass calculations for different stages of growth are based on the mass determined for the largest femur. This raises the question about what adult size is: Is it the age when sexual maturity is reached? Or is it the age at which maximum body size is reached? These are not necessarily the same. For example, many birds reach a maximum body size well before they are sexually mature, whereas crocodilians, which have indeterminate growth, reach sexual maturity several decades before reaching their "maximum" size. Erikson and colleagues (2001) compare the derived growth rates of the dinosaurs studied with the growth rate data for extant animals determined by Case (1978). It is noteworthy that Ruben (1995) has pointed out that, in Case's (1978) study, inconsistent definitions of what constituted adult size were used for endotherms and ectotherms. This has resulted in an exaggeration of differences in growth rates of endotherms and ectotherms. This aside, growth curves for extant birds, reptiles, fish, marsupials, and eutherians are based on large data sets, whereas Erickson and colleagues present data for only six dinosaur taxa, and these are of different taxonomic affinities, grossly different body sizes, and overall morphology (bipedal vs. quadrupedal). From these six representatives, they purport to deduce a generalized "dinosaurian growth pattern."

In stark contrast, studies of bone histology clearly show that different dinosaurs had very different developmental strategies and growth patterns. Surely, this direct evidence from bone should be considered in trying to assess growth rates of dinosaurs. The growth rates calculated use data from only dinosaurs that show a stratified cortex (either with zonal or vascular cycles present). For dinosaurs that do not grow in this periodic manner, it implies that their growth rates were faster. Yet these observable direct differences in growth strategy among the Dinosauria are not considered, and the derived regression line deduced is considered to be representative of the Dinosauria (Erickson et al. 2001).

The Erickson and colleagues (2001) results suggest that dinosaurs generally grew at higher rates than extant reptiles, although not as fast as altricial birds, but without considering a correction for body size differences. At a general level, these results are not in conflict with estimates from other sources, and perhaps the DME methodology may become more useful with some refinement and attention to the issues raised above. For now, on the basis of bone microstructure studies over the past several decades, it is safe to say that dinosaurs did not grow as fast as mammals and birds, but seemed to occupy a wide range of intermediate rates of growth between those of extant reptiles and birds (e.g., Chinsamy and Dodson 1995; Fig. 5.7).

Age Assessments Using Bone Depositional Rates

When growth rings occur in the primary compact bone of dinosaurs, skeletochronology can be applied, which permits deductions regarding individual age, longevity, and and so on. However, in the absence of growth rings, age assessments are much more difficult (Chinsamy 1995b), and as I have mentioned, Robin Reid suggested the use of laminae formed per year, although he cautioned that this would require several assumptions (see Chinsamy 1995b).

Recently, Martin Sander and Christian Tückmantel (2003) attempted to calculate directly the rate at which bone was being deposited in sauropod dinosaurs, by dividing the thickness of the bone between consecutive polish lines by 365 days (year) to obtain annual bone depositional rates. They then obtained mean values (although they admit to omitting certain selective values in the calculation of the mean). Bone depositional averages obtained for the four dinosaur taxa *Barosaurus* (NHUB R. 2625), *Barosaurus* (NHUB NW4), *Janenschia* (Nr 22), and *Apatosaurus* (SMA "Jacques") are 5.2, 5.1, 4.7, and 3.9 microns per day, respectively. These numbers look very similar, but in reality, the variation in the appositional rates between consecutive polish lines is quite substantial (i.e., from as low as 0.9 to as high as 20.7 microns per day). The averages above were used to calculate a mean minimal appositional rate, which was then used to estimate the maximum age of all the Tendaguru sauropods, particularly those that lacked growth lines. In addition, an intermediate and maximum age were also calculated using different averages derived from the data.

From the results, it was deduced that the smallest *Barosaurus* humerus measuring about 43.5 cm in length was probably 7–13 years old, while the longest humerus at 99 cm was probably 16–30 years old.

Although these results provide a more reasonable assessment of ontogenetic age of sauropod, it must be borne in mind that they are estimates based on a number of assumptions. For example, the thickness of the bone between consecutive polish lines is divided by 365 days, but it is more likely that in the Mesozoic a year would have been longer. Furthermore, the polish lines themselves are not accounted for in the year (i.e., how long did growth slow down/stop for in the year), and as in the earlier studies the cortical thickness of the radius is used to establish how long it took for the bone to be formed. Also when one looks at the age estimates on the basis of the number of polish lines and those calculated from the appositional rates, there appear to be fairly large discrepancies. I agree with Sander and Tückmantel (2003) that these calculations can only be regarded as "rough estimates and not as accurate calculations." Although the actual longevity data are not as accurate as we would like, it does seem to suggest that sauropods grew at faster rates than previously assumed by Case (1978) but not as fast as Currey (1999) deduced for *Apatosaurus.*

Conclusions

The preservation of the microscopic structure of dinosaur bone permits a host of deductions about the various factors that affected the growth of the bone. Detailed analyses of changes in bone microstructure during ontogeny of several dinosaur taxa have permitted a good understanding of the nature of bone tissue types that characterizes various stages of ontogeny and permit qualitative deductions about associated levels of growth. For example, as in modern vertebrates, dinosaurs grew rapidly early in ontogeny and had slower growth rates during later ontogeny. In many of the dinosaurs studied, we see evidence of slowed or even determinate growth patterns, while others seem to have an indeterminate growth strategy. Studies of several dinosaur taxa suggest that many dinosaurs still retain the plesiomorphic flexible growth strategies (as reflected by the occurrence of zonal bone), although some dinosaurs appear to be able to grow at sustained rapid rates for most of their lives. In general, however, it appears that dinosaurs were incapable of very rapid growth rates typical of extant birds and mammals.

When growth marks are present (i.e., LAGs and/or annuli and polish lines), skeletochronology have been applied (Chinsamy 1990, 1993a; Varricchio 1993; Erickson

1999, 2001) to assess individual "age" and derive growth trajectories. In the absence of periodic growth marks, or whether such growth marks are not considered to be annual features, the thickness of the cortical bone, divided by a bone depositional rate, have been used to obtain "age" information about the dinosaur. The bone depositional rates are either extrapolated from modern vertebrates to dinosaurs (e.g., Horner et al. 2000) or derived by measuring laminae thickness between consecutive growth marks in some dinosaur taxa and applying this to other taxa without growth marks (Sander and Tückmantel 2003). Table 5.1 shows age data for various dinosaur taxa.

Application of each of the above methodologies appears not to be without problems. Skeletochronology depends on the assumption that the growth rings are annually formed (as in extant reptiles) and, in many cases, requires back-calculations to account for growth rings obliterated by normal remodeling processes of growth. Because of variability in the skeleton, choice of the skeletal elements used for skeletochronology is crucial. With regard to the method of extrapolating bone depositional rates of modern taxa to fossil taxa, as discussed at length previously, similar bone tissue types do not imply similar bone depositional rates and invoking this actualistic hypothesis is inappropriate. Furthermore, given the variability in bone depositional rates of extant vertebrates and even for a particular tissue type within a particular clade of birds, making such extrapolations are undermined.

In some ways, bone depositional rates derived from the bone itself are better. The assumptions here are that the growth marks are annual, and a year in the Mesozoic is equivalent to 365 days and no consideration of the "duration" of "polish lines" is made. Furthermore, the use of an average bone depositional rate throughout ontogeny fails to consider ontogenetic variation in bone depositional rates as well as variation in the rate of bone deposition because of the local morphology of the bone. Linear radial appositional rates are also measured; perhaps a better measure would be a consideration of the total area of bone formed because this would also consider the entire cross section of the bone. This methodology also measures only radial cortical thickness of the bone without accounting for bone removed at earlier stages of growth.

Careful analyses of the various growth series studies of dinosaurs suggest that dinosaurs did not grow as fast as birds, nor did they grow as slowly as extant reptiles. Dinosaurs appear to have had variable growth rates that were intermediate between that of extant endotherms and ectotherms (Fig. 5.7B).

Table 5.1
Age Estimates to Adult Size Based on Growth Ring Counts, Polish Lines, and Appositional Rates

			Estimated age (years)		
Taxon	Reference	Estimated adult body mass (kg)	Growth rings	Polish lines	Appositional rates (min.; intermed.; max.)
Shuvuuia	6	1.7	3		
Psittacosaurus	5	20	9		
Syntarsus	1	20–25	8		
Troodon	12	50	3–5		
Massospondylus	2	250	15		
Maiasaura	8	1500	6–7		
Lapparentosaurus	4, 9	Subadult	20		
Apatosaurus	3, 11	25,000	6[a]	25	3.9; 8.2; 33.1
Janenchia[b]	10, 11			26[c]	18.6; 24.3; 35.1
Brachiosaurus[b]	11				26.8; 35.0; 50.
Barosaurus[b]	11			14	20.5; 26.8; 38.6
Dicraeosaurus[b]	11				11.4; 14.9; 21.5
Tyrannosaurus rex	7	5000	20		
Gorgosaurus, Albertosaurus, Daspletosaurus	7	750; 1000; 1500	16–20		

Note: References cited are as follows: 1, Chinsamy 1990; 2, Chinsamy 1993a; 3, Currey 1999; 4, de Ricqlès 1983; 5, Erickson and Tumanova 2000; 6, Erickson et al. 2001; 7, Erickson et al. 2004; 8, Horner et al. 2001; 9, Rimblot-Baly et al. 1995; 10, Saunder 1999; 11, Sander and Tückmantel 2003; 12, Varricchio 1993.
[a]Currey 1999.
[b]Largest specimen of sample studied.
[c]Sander 1999.

CHAPTER SIX

Biology of Early Birds

Modern birds are an extremely diverse group represented by about 10,000 species. Yet these modern lineages (the Neornithes) that we are so familiar with, represent just one branch on the evolutionary tree of birds. Molecular sequence data suggest that neornithine lineages originated some 90–100 million years ago in the Cretaceous. In a fascinating study of phylogenetic relationships and historical biogeography of basal Neornithes lineages, Joel Cracraft (2001) of the American Museum of Natural History has proposed a Gondwanan origin of neornithines before the Cretaceous-Tertiary extinction event. This is, however, unsubstantiated by the fossil record, and the sparse reports of supposedly Cretaceous neornithine birds are quite controversial. The earliest uncontroversial neornithine birds date to about 55 million years ago in the Early Eocene. Perhaps new fossil discoveries will shed more light onto the question of when, how, and where, modern birds originated. It is interesting that new discoveries of a host of fossil birds close to the very beginning of birds (i.e., all birds, not just the Neornithes) has intensified research into the origins and early radiation of birds, and there is now much more focus on the biology of these early birds.

The evolutionary history of birds dates to about 150 million years ago, to a time when dinosaurs dominated the earth and pterosaurs soared in the air. The first clue of birds in the fossil record came in 1860, with the discovery of a single feather in the

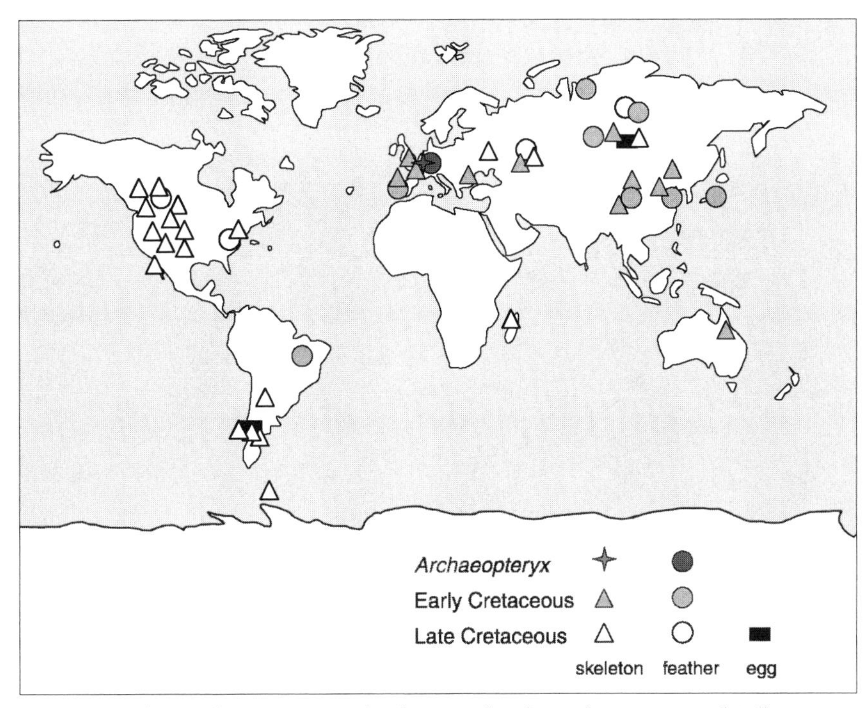

Fig. 6.1. World map of Mesozoic avian biodiversity (data from Chiappe 1995 and Kellner 2001).

fine-grained limestones of Solenhofen, Germany. The next year the enigmatic skeletal remains of *Archaeopteryx lithographica* that preserved feathers on its wings and on its long, bony tail were discovered. Today, seven skeletons and the original feather (all from the same locality) represent *Archaeopteryx*.

Ever since the discovery of *Archaeopteryx,* the question of its ancestry has concerned many because, even though it was feathered, in anatomical structure, it appears to be transitional, between birds and reptiles. For more than the 140 years since its discovery, scientists have grappled with the dichotomy of characters represented in *Archaeopteryx,* and a wide range of hypotheses exists regarding both its ancestry and the evolution of flight. In recent years, several reviews have been published about the possible ancestors of birds (e.g., Padian and Chiappe 1998; Witmer 2002); therefore, I will not delve into this. Today, most paleontologists endorse the dinosaurian hypothesis first proposed by Thomas Huxley in the 1860s and greatly expanded and championed by John Ostrom of Yale University in the 1970s. These days the debates tend not to be centered on whether birds are dinosaur descendants but rather about which dinosaurs represent the sister taxon of birds (i.e., whether the dromaeosaurs, troodontids, oviraptorids, or alvarezsaurids are most closely related to birds).

Synopsis of Mesozoic Avian Diversity

Our knowledge and understanding of the diversity and evolution of early birds have advanced over the past few years. This is directly attributable to the surge in fossil birds discovered around the world, particularly from China (Fig. 6.1). It is no exaggeration that the number of Mesozoic birds discovered in just the past 10 years more than trebled all those known before. The explosion of discoveries of Mesozoic birds has provided new insights into the early radiation of birds and the evolution of flight, as well as elucidated their phylogenetic relationships.

Even though many more Mesozoic birds are now known, none of these predate the 150-million-year-old *Archaeopteryx.* In the 1980s, *Protoavis* from the Dockum Group of Texas, about 75 million years older than *Archaeopteryx,* was proposed by Sankar Chatterjee as the ancestor of birds. However, the two disarticulated remains of *Protoavis* from the Post Quarry are rather poorly preserved, and the newer finds from the Kirkpatrick Quarry, though assigned to *Protoavis,* may in fact represent enigmatic new taxa. The problem, however, is that this new material, consisting of mostly postcranial bones from the older, Late Carnian Tecovas Formation is disassociated, and according to Larry Witmer (2002), it may represent a composite of different taxa. My recent work with Andrzej Elzanowski and Chatterjee suggests that these Tecovas femora are indeed taxonomically distinct from the *Protoavis* holotype femur (A. Chinsamy, A. Elzanowski, and S. Chatterjee, unpublished manuscript).

In 1995, Korean researchers reported Late Jurassic birds from China and North Korea, but the problem here is that the actual age of these deposits is controversial. The North Korean specimen dubbed the "North Korean *Archaeopteryx*" is anatomically quite different from *Archaeopteryx. Confuciusornis* from Yixian Formation of northeastern China has also been publicized as a Late Jurassic bird, predating *Archaeopteryx,* but this appears to be confounded because of inaccurate dating. Recent palynological studies have suggested a lowermost Cretaceous age for these deposits, and even more recent laser-derived 40Ar-39Ar dating have dated the deposits to about 121–122 million years ago. Thus, the 150-million-year-old *Archaeopteryx* is still the undisputed oldest fossil bird.

Despite more than a century of study, various aspects of *Archaeopteryx* biology remain unresolved. At the recent Society of Avian Paleontology and Evolution meeting in Beijing, China, it was evident that the debate still rages on about a number of aspects about the lifestyle and biology of *Archaeopteryx.* For example, there is still no consensus about whether *Archaeopteryx* was cursorial or arboreal, whether it had a perching foot, whether it was a glider or capable of flapping flight, or whether they were endothermic like modern birds. In fact, even though most people support the dinosaurian origin of birds, it was obvious at this meeting that there were a

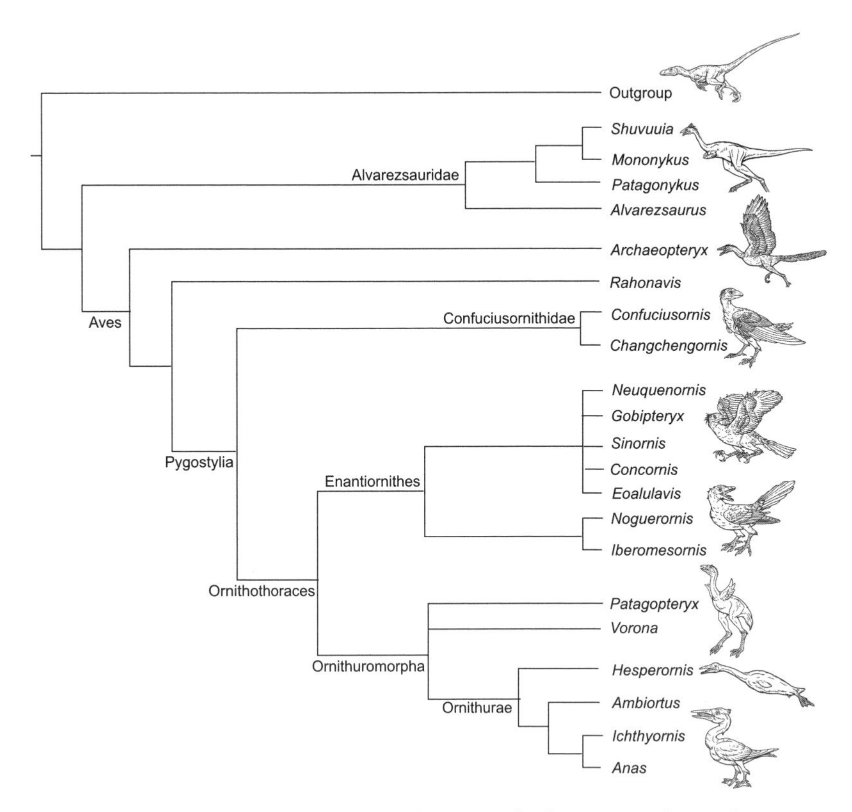

Fig. 6.2. Cladogram of relationships of Mesozoic bird. Courtesy of Luis Chiappe.

small, vocal group of scientists who supported alternative hypotheses for the evolution of birds.

The early history of birds shows that not long after *Archaeopteryx* a number of diverse forms evolved (Fig. 6.2). Anatomical studies of some of these birds have shed light onto their phylogenetic relationships and also on various aspects of their biology, especially pertaining to their behavior and lifestyles. For example, the Early Cretaceous *Sinornis* from China (Sereno and Rao 1992) and *Iberomesornis* and *Eoalulavis* from Spain (see Chiappe and Witmer 2002), show adaptations for enhanced flight capabilities. By the Late Cretaceous, *Ichthyornis* shows advanced flight capabilities, while *Hesperornis* from contemporaneous deposits shows striking adaptations for diving. Flightless birds in the Cretaceous are also well represented (e.g., *Patagopteryx*), while by far the most widespread and diversified Cretaceous birds are the volant enantiornithines. At present, twenty species of enantiornithine birds are recognized, and this number is set to increase over the next few years.

One of the fascinating new fossil bird discoveries is that of *Jeholornis* from western Liaoning in northeastern China (Zhou and Zhang 2003), which, except for *Archaeopteryx,* is the only known bird with a complete, long skeletal tail. In fact, its tail is even longer than *Archaeopteryx,* and its tail feathers are more like those of dromaeosaurs than *Archaeopteryx.* On the basis of these finds, Zhonghe Zhou and Fucheng Zhang have proposed that there must have been a more primitive bird before *Archaeopteryx.* It is interesting that the discovery of fifty different types of seeds in the stomach region of *Jeholornis* suggests that already in the Mesozoic seed-eating adaptations had evolved in birds.

Feathered Dinosaurs

Until 1996, there was no reason to suspect that dinosaurs were covered by anything other than reptilelike scales. Bob Bakker's (1975) reconstruction of a feathered *Syntarsus* with a tuft of feathers on its head was completely speculative. However, at the 1996 Society of Vertebrate Paleontology Meeting in New York, news leaked out about the discovery of a feathered dinosaur from northeastern China. A photograph that circulated showed distinct featherlike structures all along the dorsal part of the animal. This find astonished the paleontological gathering. For the first time, a dinosaur showed something other then the typical reptilian-like integument.

Fossilized feathers were not new to the paleontological community. As early as 1861, from the Solnhofen Lithographic Limestones of southern Germany, an almost complete fossilized primary flight feather was discovered and named *Archaeopteryx* ("*ancient wing*"). Several of the referred skeletal material later discovered from these fine-grained limestones preserved impressions of feathers, and several other isolated feathers attributed to *Archaeopteryx* are known. Therefore, it was not just the featherlike structures that startled the paleontological community but rather the fact that until the discovery of *Sinosauropteryx,* as the Chinese dinosaur was later named, all feathered creatures in the fossil record were birds (Fig. 6.3). This find of a nonavian dinosaur with featherlike structures rekindled the debate about the origin of birds and reinforced the hypothesis for the origin of birds from dinosaurs.

A team of paleontologists consisting of Larry Martin, Peter Wellnhofer, and John Ostrom went to China to investigate the so-called feathers of this unique dinosaur. After close scrutiny, they announced that the integumentary structures on *Sinosauropteryx* were "protofeathers," which were possibly evolutionary forerunners of feathers. In the formal publication of the specimen (Chen et al. 1998), besides the feathery integument, it was reported that oviducts and ova were also preserved within the body cavity of the specimen. A year later, a second specimen of *Sinosauropteryx* was discovered that also

Fig. 6.3. *Sinosauropteryx,* the first of the "feathered" dinosaurs discovered from the Liaoning locality in northeastern China. The featherlike structures are located all along the dorsal (top) part of the skeleton (arrow). Body length is 70 cm. Image courtesy of Willem Hillenius.

Fig. 6.4. *Caudipteryx,* the 91 cm tall, feathered dinosaur from Liaoning in northeastern China. Although flightless, the plume of feathers (arrow) on its tail and on its forelimbs show the typical asymmetric design as in modern bird feathers. Image courtesy of Zhonghe Zhou.

preserved a "featherlike" integument and the remains of its last meal (a mammal) within its body cavity. Most paleontologists agree with the interpretation of the dorsal midline structures in *Sinosauropteryx* as protofeathers, although there are some who have offered counterarguments. For example, Nick Geist, Terry Jones, and John Ruben (1997), in dissections of squamates such as *Varanus,* found that macerated collagen fibers looked very much like the structures on *Sinosauropteryx,* and suggest that the fossilized structures should therefore not unequivocally be regarded as protofeathers.

Two years later, on June 23, 1998, at a press conference at the National Geographic Society in Washington D.C., Ji Qiang and Ji Shu-An of the National Geological Museum in Beijing; Phil Currie of the Royal Tyrrell Museum of Palaeontology in Drumheller, Alberta; and Mark Norell of the American Museum of Natural History in

New York announced the discovery of two more feathered dinosaurs in Liaoning Province in China: *Protarchaeopteryx robusta* and *Caudipteryx zoui* (Ji et al. 1998; Fig. 6.4). Although this time there is no doubt that the structures are feathers, both these animals are considered to be incapable of flight because of their fairly small arms. The 91 cm tall *Caudipteryx* preserves feathers on its arms and tail (Fig. 6.4). The turkey-sized *Protarchaeopteryx* has symmetrical feathers, that covered its body, but no wing feathers are preserved. Although both animals are flightless, true flight feathers are present, which suggest that their ancestors probably flew. *Caudipteryx* is distinctive in having a fan of tail feathers, which were probably used for display (Fig. 6.4). Some researchers agree that the feathers on *Protarchaeopteryx* and *Caudipteryx* are consistent with modern bird feathers, but they suggest that these dinosaurs are ambiguously recognized as nonavian, and that they are rather in fact birds (e.g., Jones et al. 2000). Assessing cursoriality in bipedal archosaurs, Terry Jones from Oregon State University and colleagues (2000) found that *Caudipteryx* had a running mechanism (i.e., with body mass centered interiorly, near the wings) as in cursorial birds. On the basis of these findings, they propose that *Caudipteryx* could have been a secondarily flightless bird and not a nonavian theropod dinosaur. However, Luis Chiappe and several other researchers, such as Paul Sereno, Tom Holtz, Mark Norell, and Phil Currie (see Chiappe and Witmer 2002), are confident that cladistic analysis firmly recognizes *Caudipteryx* as a nonavian dinosaur, outside of Aves.

Today there are eight species (six genera) of dinosaurs from the Liaoning fossiliferous deposits that preserve feathers, although there are apparently dozens more that are yet to be described. In January 2003, the most recently described spectacular-feathered dinosaur, *Microraptor gui,* made the headlines (Xu et al. 2003; Fig. 6.5). (A section on *Microraptor* follows.)

The fascinating feathered dinosaurs show a diverse group of animals with a variety of different feather structures: *Sinosauropteryx,* the first "feathered" dinosaur described with the simple filamentous-like structures; *Beipiaosaurus,* a therizinosauroid shows filamentous structures as in *Sinosauropteryx; Sinornithosaurus,* a dromeaosaurid has tufts joined at their bases or serially arranged along a central filament; and then *Caudipteryx* (Fig. 6.4), *Protarchaeopteryx,* and *Microraptor* (Fig. 6.5) with complex feathers with vanes and shafts.

Richard Prum (1999) has recently proposed a developmental and evolutionary model that explains the homology of these varying feather types with that of modern bird feathers. (At the 2004 International Comparative Vertebrate Morphology Congress in Florida and at the Society of Avian Paleontology Meeting in France, Paul Maderson contested Prum's hypothesis and proposed an alternative model for the development of feathers that suggests a homology of the integumentary structures of *Longisquama* with avian

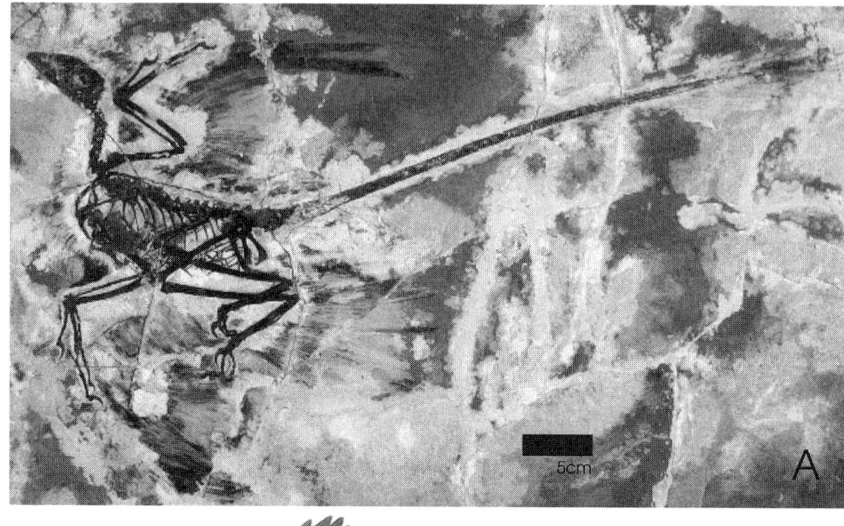

Fig. 6.5. *A. Microraptor gui,* the four-winged dinosaur from Liaoning in northeastern China. *B.* Reconstruction of *Microraptor gui.* Image courtesy of Zhonghe Zhou.

feathers; see Maderson and Hillenius 2004.) Prum has proposed that feathers evolved through five stages: (1) the evolution of a hollow elongated tube, (2) a downy tuft of barbs, (3) a pennaceous structure, (4) barbules and hooklets evolved to create a closed-vaned pennaceous feather. (5) Asymmetrical vanes of flight feathers would have been the last step in the evolutionary development of feathers. Support for this theory comes from the diversity of feather types among modern birds, and from molecular data, as well as from the range of feathers that are preserved on the dinosaurs from Liaoning. The most primitive feathers, which represent stage 1 in Prum's evolutionary model, are those found on *Sinosauropteryx*. Increasingly more complex feathers are easily recognizable in the variety of feathered dinosaurs discovered at Liaoning.

The evidence from the fossil record suggests that flight was not the driving force for the evolution of feathers. Thus, feathers may have initially evolved for other purposes, such as insulation, display, or camouflage. Camouflage and display are opposite ends of the spectrum of visibility. Camouflage would mean that the feather color would blend in with the habitat of the bird, while as in modern birds, display would require feathers to be conspicuous and brightly colored. Whether feathers evolved for insulation is uncertain. However, the fact that they were present might have afforded insulation. It is interesting that, with the trend to small size in the theropod lineage leading to dinosaurs, there appears to be a concomitant complexity of feather development.

Even before the discovery of the exquisite feathered dinosaurs of Liaoning, the suite of modern feathers present in *Archaeopteryx* seemed to suggest that the origin of feathers preceded the origin of birds. The feathered dinosaurs from Liaoning support this hypothesis and show undisputedly that feathers evolved in a lineage of theropod dinosaurs even before the origin of modern avian flight.

Microraptor

In January 2003, the startling discovery of the dromaeosaur *Microraptor gui* was announced to the scientific community (Xu et al. 2003). By now, everyone was aware of the feathered dinosaurs from Liaoning, so having feathers was not enough to command the attention of the world. This was not just another feathered dinosaur: *Microraptor* is unique in that it has asymmetrical feathers on both its forelimbs *and* hindlimbs (Fig. 6.5).

Until now, the origin of flight had mainly been debated from views that either suggested that flight evolved from the "ground upward," (i.e., arboreal theory) or from "trees downward" (i.e., the cursorial hypothesis). Until *Microraptor*'s discovery, the feathered dinosaurs from Liaoning seemed to support the ground upward hypothesis, which suggests that flight evolved from a cursorial theropod. However, the four-winged *Microraptor* suggests that modern avian flight could have evolved from such an ancestor and supports the arboreal hypothesis for the origin of flight. It is ironic that, as early as 1915, William Beebe had proposed that avian flight evolved through a four-winged, tetrapteryx stage, but until now, his work had not received much attention.

The presence of asymmetrical feathers on both the fore- and hindlimbs suggest aerodynamic capability, and the researchers Xu and colleagues (2003) propose that the wings were used for gliding. They further propose that with the evolution of powered flight in the forelimb, the hindwings became increasingly reduced and were later lost in birds.

Even before *Microraptor* was described, Mark Norell and colleagues (2002) described vaned feathers on the hindlimbs of an unnamed dromaeosaur. It is possible that the fore- and hindwings are unique to the dromaeosaurs, but for now, researchers are grappling with how these dinosaurs used their four wings and whether the forewing is homologous to that of *Archaeopteryx* and modern birds. Questions persist about whether the arboreal hypothesis for the origin of flight is also consistent with the origin of birds from dinosaurs. In the past, the dinosaurian hypothesis for the origin of birds was strongly linked to the cursorial—ground upward hypothesis for the origin of flight. However, given that undisputed nonavian dinosaurs such as the dromaeosaurs were feathered and possibly flew, surely, the arboreal model cannot be considered inconsistent with the dinosaurian origin of birds.

Bone Microstructure of Mesozoic Birds

Studies of the overall morphology and anatomy of Mesozoic birds permit deductions about their phylogenetic relationship, as well as deductions about the biomechanics of the skeleton and adaptations for particular lifestyles. An assessment of the internal geometry and relative bone wall thickness (RBT) of the bones infers the biomechanical properties of the skeleton, while studies of the internal organization of the bone (i.e., its bone microstructure) are increasingly recognized as providing direct evidence about various aspects of their biology. For example, the texture of the bone microstructure permits direct deductions regarding the qualitative rate at which bone formed (i.e., fast or slow), as well as the pattern of bone formation (i.e., whether bone deposition was continuous or periodically interrupted). However, perhaps because of the scarcity of bird fossils, studies on the bone microstructure of basal birds lagged far behind that of nonavian dinosaurs.

In 1987, Peter Houde had undertaken studies describing the bone microstructure of the Cretaceous ornithurine bird, *Hesperornis;* but until 1994, no other studies on the microscopic structure of Mesozoic birds were done. In 1994, my colleagues and I published our findings on the bone microstructure of three nonornithurine bird taxa (Chinsamy et al. 1994). Of the twelve Mesozoic bird taxa whose bones have been studied histologically, my colleagues and I have described the bone microstructure of eleven of them. Framed within the context of ontogeny and phylogeny, our studies have provided valuable insights into the biology of basal birds.

Bone Microstructure of Nonornithurine Birds

My work on the microscopic structure of bones of early birds began in 1993, soon after I arrived in Philadelphia, Pennsylvania, for postdoctoral research with Peter Dodson at the University of Pennsylvania. About the same time, Luis Chiappe had also begun his postdoctoral work at the American Museum of Natural History in New York. Chiappe and I had met the summer before at a Society of Avian Paleontology and Evolution meeting in Frankfurt and while there began to plan our collaborative study on the bone microstructure of basal birds. We began our studies with three basal birds that were recovered from the Cretaceous deposits of Argentina. Through the generosity of G. Rougier (formerly of the Museo Argentino de Ciencias Naturales, Buenos Aires) we obtained a specimen of the hen-sized flightless bird, *Patagopteryx deferrariisi*. *Patagopteryx* was found in the Coniacian-Santonian Rio Colorado Formation of southern Argentina and is regarded as the sister group of the Ornithurae (which includes hesperornithiforms, ichthyornithiforms; and modern birds; Fig. 6.2). Included in the assessment were also femora of two distinct enantiornithine bird taxa from the Maastrichian Lecho Formation of northwestern Argentina. (Because the morphology of these bones is described in detail in Chinsamy et al. 1995, we will not go into this here.) The enantiornithines represent a diverse group of birds that were capable of flight and that had a worldwide distribution. Their relationship to other avian lineages has been fairly controversial, although recent cladistic analyses place them as intermediate between *Archaeopteryx* and the ornithurine birds (Chiappe 1995; Fig. 6.2).

Until this research, my experience with bone microstructure of birds was restricted to that of modern taxa. As part of my PhD, I had undertaken an assessment of the histological changes through ontogeny in the ostrich *Struthio camelus* and the secretary bird *Sagittarius serpentarius* (e.g., Chinsamy 1995a). Thus, when I first looked at the thin sections of the femur of *Patagopteryx* under the microscope at the petrographic laboratory at Bryn Mawr College (near Philadelphia), I was immediately struck by how different its microstructure was from that of modern bird bones.

Like modern birds the bone wall consisted of a richly vascularized fibrolamellar bone tissue (Plate 6). The medullary cavity was lined with a circumferential layer of parallel-fibered bone and a distinct tide line marked the endosteal nature of this bone tissue and highlighted the fact that bone resorption had occurred before the deposition of this perimedullary bone tissue (Plate 6). The bone tissue differed from that of modern birds in several ways: in overall arrangement, it lacked the typical triple-layer structure generally associated with adult modern birds (i.e., with an ICL and OCL of bone surrounding a central tissue of fibrolamellar bone (e.g., Enlow and Brown 1957; Chinsamy 1995a; see Fig. 3.11A). The orientation of the channels in which the vascular canals were located was also mainly longitudinally oriented, with some circumferential and radial arrangements unlike the more usual reticular arrangement generally seen in modern birds. However, the most striking characteristic of *Patagopteryx* bone microstructure was the presence of a distinct line of arrested growth (LAG) and an annulus that could be traced around the entire compacted bone wall, which interrupted the deposition of fibrolamellar bone (Plate 6; Fig. 6.6). As mentioned before, the presence of a LAG directly indicates an arrest of growth, while the annulus that preceded its deposition directly suggests a slower rate of bone deposition as compared with the fibrolamellar bone tissue. After the period of slowed growth, and an arrest in bone deposition, *Patagopteryx* resumed rapid growth as is evidenced by the fibrolamellar bone tissue. I immediately recognized the bone tissue was similar to that of several nonavian theropods I had previously examined and quite unlike the bone tissue of modern birds that generally grow in a rapid, sustained, uninterrupted manner (cf. modern bone tissue in Fig. 3.11 and that of nonavian dinosaurs, e.g., Fig. 4.3).

Transverse sections of the femora of the two enantiornithine birds revealed very thin walled bones, with a relative bone wall thickness of only about 13.7% (as compared with the flightless *Patagopteryx* with RBT of 18%) and a large free medullary cavity. Examination of the bone microstructure of the two enantiornithine femora under the microscope revealed that they were even more unusual than modern birds, as compared with *Patagopteryx*. The strikingly distinctive characteristics of the enantiornithine bone microstructure is that the relatively thin bone wall consisted almost entirely of poorly vascularized, slowly formed parallel-fibered bone that was periodically interrupted by several lines of arrested growth (LAGs; Fig. 6.6). Five LAGs were present in the cross section of PVL-4273, and in MACN-S-01, four LAGs were counted (e.g., Chinsamy et al. 1994). The medullary cavity in both enantiornithines is lined with a thin deposit of endosteally derived lamellated bone tissue. The lack of fibrolamellar bone in the enantiornithines is quite unusual, and although the tide line that separates the endosteal bone suggests that earlier bone tissue was resorbed during remodeling processes, whether this earlier bone tissue was more rapidly formed could not be ascertained. Therefore, although, the parallel-fibered nature of the bone wall is similar to the OCL of modern birds (which is deposited when they have almost reached adult size), it is distinctive in the enantiornithines in that it makes up the entire bone wall, it has growth rings, and it is also substantially thicker (Chinsamy and Elzanowski 2001; see Figs. 3.11, 6.6).

The poorly vascularized nature of the enantiornithine compacta is also worth noting. In some areas of the bone wall, enlarged osteocyte lacunae with hyperramified canaliculi (EOL/HRC) tend to occur (Fig. 6.6). Such structures are not evident in the bones of modern birds, although in a recent study, I have observed similar EOL/HRC structures in the Tecovas femora (incorrectly assigned to *Protoavis*) from the Kirkpatrick Quarry. My collaborators and I suggest that these EOL/HRC may have facilitated the assimilation and distribution of nutrients (as well as the removal of waste products) in the rather poorly vascularized bone compacta (e.g., Chinsamy et al. 1994).

The presence of LAGs in all three Cretaceous birds indicates that they periodically experienced a pause in the rate at which they grew, and because these are not preceded or followed by an annulus (as in, e.g., *Patagopteryx*), it further suggests that such arrests in growth were fairly abrupt. This cyclical pattern of growth implies that these birds grew at much slower rates as compared with modern birds (and to *Patagopteryx*). In addition, considering that such LAGs are generally annually formed in extant vertebrates, it suggests that these enantiornithines and *Patagopteryx* required more than 1 year to reach a mature body size (Chinsamy et al. 1994). This suggests a fundamental difference in the growth strategy of basal birds as compared with modern birds, which generally grow to adult size within a year.

Before our study, the only other Mesozoic bird studied histologically was *Hesperornis* (an ornithurine diving bird), but its histology appeared to be similar to that of modern birds. Our study of *Patagopteryx* and the enantiornithines represented the most primitive birds ever examined and our findings suggested that these birds grew much more slowly as compared with modern birds (e.g., Chinsamy et al. 1994; Chinsamy 2002). Because these birds were more derived than *Archaeopteryx,* it was quite likely that even *Archaeopteryx* grew at similar slow, interrupted rates, quite unlike modern birds, which typically grow to adult size within a single year (Chinsamy and Dodson 1995). Our hypothesis is supported by the fact that the seven skeletons of *Archaeopteryx* represent a substantial size range; all of them have limited skeletal fusions, which suggests that they are all subadults, the youngest at about 50% the size of the largest (Chiappe and Dyke 2002). Further support of our hypothesis of slowed growth in basal birds is that Chiappe et al. (1999) have recently reported that in the growth series of *Confuciusornis sanctus,* individuals lacking neonate features are differently sized, with the smallest specimens about 50% and 60% the size of the largest. Chiappe et al. (1999) mentions that ornithologists working on modern birds would possibly argue that size differences of such magnitudes are suggestive of different species; however, such a wide range of sizes appears to be evident in growth series of both *Archaeopteryx* and *Confuciusornis.* These independent findings reinforce our hypothesis that basal birds had slower rates of growth than their extant relatives (Chinsamy et al. 1994; Chinsamy and Dodson 1995; Chinsamy 2002).

Evidence for slowed growth in *Confuciusornis* came from independent studies of its bone microstructure. The Late Jurassic–Early Cretaceous *Confuciusornis* from Yixian Formation of Western Liaoning was first studied by Zhang Fucheng and colleagues (1998, 1999) and more recently by de Ricqlès and colleagues (2003). The initial studies were conducted using scanning electron microscopy, which makes comparisons with the thin-sectioned material of other Mesozoic birds rather difficult. However, Zhang and colleagues mention that the structure differs from *Patagopteryx* and the enantiornithines and is more similar to modern birds (the marine birds *Larus ridibundus* and *Puffinus leucomelas* and a galliform from Zhoukoudian) included in their study. They report that the compacta of *Confuciusornis* consists of the triple-layer structure typical of modern birds, although the bone wall is thicker than that of the modern birds, which they interpret as suggesting weaker flying abilities. Zhang and colleagues (1998) also mention the presence of rest lines in the OCL of bone. The recent comprehensive assessment of the bone microstructure of *Confuciusornis sanctus* by de Ricqlès and colleagues (2003) concurs with the earlier reports made by Zhang and colleagues (1998, 2000). Here, the femoral bone wall thickness is given as 25%, which is higher than for other flying Mesozoic birds and, curiously, higher than for the flightless *Patagopteryx*. As in other nonornithurine birds, a tibial section shows a LAG outside the OCL. In this publication, researchers assume a depositional rate of at least 10 microns per day on the basis of extrapolations from living birds and deduce that the diameter of the bone could have been constructed in 24 weeks or less (Ricqlès et al. 2003). I do not agree with the calculations for a number of reasons: the researchers assume constant radial appositional bone depositional rates, when in reality bones do not grow like this; they do not account for resorption of bone from the perimedullary region; no consideration is made of time lapsed during LAG formation; and most important, assuming a bone depositional rate of 10 microns per day fails to acknowledge that bone depositional rates (of even similar tissue types) are quite variable among extant taxa, even within the clade of modern birds. Furthermore, bone depositional rates are known to vary according to ontogenetic age and local conditions of growth. Notwithstanding their methodology, these researchers conclude, as we did (Chinsamy et al. 1994), that basal birds grew more slowly than extant birds.

The Malagasy Birds

More details of the growth patterns of Mesozoic birds became evident as this research expanded to include three more Late Cretaceous, nonornithurine birds from the Maevarano Formation of the Majunga Basin of northwestern Madagascar (Rogers et al. 2000). The first of these pre-Holocene Malagasy birds, *Vorona berivotrensis,* was recovered in 1995 near the village of Berivotra and described later by Cathy Forster and colleagues (1996). Because of *Vorona*'s relatively incomplete nature, its relationship to other nonornithurine basal birds is still not completely resolved. Recent cladistic analyses place *Vorona* as separate from the enantiornithines and closer to *Patagopteryx* and Ornithurae (Figs. 6.2, 6.6). The second Malagasy bird studied histologically was *Rahonavis ostromi,* which is considered to be (after *Archaeopteryx)* one of the most basal birds. Cladistic analyses place it as a sister taxon to *Archaeopteryx.* The third Malagasy bird worked on histologically is a yet unnamed Malagasy enantiornithine.

All three nonornithurine Malagasy birds showed well-preserved bone microstructure with periodic interruptions in the rate of bone formation (Fig. 6.6). This condition is similar to that described earlier for the nonornithurine birds, *Patagopteryx* and two enantiornithines.

Embryonic Bird Bone Histology

Embryonic bones of early birds are very rare. Therefore, the discovery of seven specimens of embryonic *Gobipteryx* (Fig. 6.7), including egg shell fragments, from the red beds of Barun Goyot Formation at Khermeen Tsav in the Gobi Desert caused a stir in paleontological circles. Polish paleontologist Andrzej Elzanowski described these exquisite specimens in 1981, and he and I recently described the bone histology of these embryos (Chinsamy and Elzanowski 2001; Fig. 6.7).

We found that, like most embryonic bone, the bone consisted of a woven bone matrix with a large number of cancellous spaces that entrapped blood vessels. No centripetal deposition of osteonal bone had occurred within these spaces, and osteocyte lacunae appeared to be large, globular structures that were fairly randomly distributed in the woven bone matrix. The uneven peripheral margins of the bone suggest that bone deposition was underway in this region at the time of death.

The two adult enantiornithine birds studied from the Maastrichtian Lecho Formation of Patagonia do not preserve this fibrolamellar type of tissue in the compacta of their bones, which suggests that this rapidly formed embryonic tissue must have been resorbed by remodeling processes during later, postembryonic growth (Chinsamy et al. 1994; Chinsamy and Elzanowski 2001).

Bone Microstructure of Cretaceous Ornithurine Birds

After the initial study of *Patagopteryx* and the two enantiornithine bird bone microstructure (Chinsamy et al. 1994, 1995), I became interested in trying to ascertain when, in the evolution of birds, did the typical rapid growth strategy of modern birds evolve. I began looking for other Cretaceous birds to study, and through the generosity of Larry Martin of the Kansas Museum, I began working on *Hesperornis, Ichthyornis,* and *Cimolopteryx* (Fig. 6.2). Two diagnostic femoral fragments of *Hesperornis* recovered from the Late Cretaceous Niobrara Chalk were examined in our study.

Hesperornis had been studied previously by Peter Houde (1987) working at the time in the National Museum of Natural History at the Smithsonian Institution in Washington, D.C. He attempted to use the bone histology to ascertain whether this foot-propelled diving bird was a paleognath or a neognath. (These terms refer to the structure of the jaws of modern birds: paleognathous birds [paleognaths], such as tinamous and ratites [e.g., rheas, ostriches] have a more loosely constructed mobile palate, while neognathous birds have a less mobile palatal structure.) Peter Houde's studies were prompted by that of the Italian researchers Amprino and Godina (1947), who suggested that paleognaths and neognathous birds showed distinct differences in the orientation of their blood vessels. They described neognathous birds as having vascular canals that were reticular, branched, and that anastomized randomly. Houde agreed with this description, but reevaluated their description of paleognathous bone microstructure. He proposed that paleognathous bird bone histology could be divided into a tinamou-type and a ratite-type, with the tinamou-type having the vascular canals longitudinally arranged and oriented parallel to the longitudinal axis of the bone. He considered the ratite-type to have a plexiform or laminar arrangement of vascular canals, arranged in closely packed concentric circles in the transverse plane of the bone. On the basis of these distinctions, Houde then described the bone of *Hesperornis* as a neognathous-type, and deduced that *Hesperornis* was therefore more closely related to neognathous birds.

My study of the bone microstructure of *Hesperornis,* and with comparison of ontogenetic changes in bone microstructure of modern birds (Chinsamy 1995a; Starck and Chinsamy 2002), showed that the above division into paleognathous and neognathous bone types is rather simplistic, especially because variation in blood vessel orientation changes during ontogeny and that localized areas in thin sections of *Hesperornis* bone can be ascribed to either of the types. For example, extant juvenile birds tend to have more longitudinally arranged blood vessels that tend to anastomize only later in ontogeny to circumferential or reticular or plexiform arrangement (Chinsamy 1995a; Starck and Chinsamy 2002). Incidentally, Padian et al. (2002) also fail to consider histological changes during ontogeny and suggest that modern birds have reticular-plexiform arrangements of vascularization. A distinctive characteristic about *Hesperornis* bone is that as in modern birds it consists essentially of a fibrolamellar type of bone tissue, with no periodic interruptions in growth,

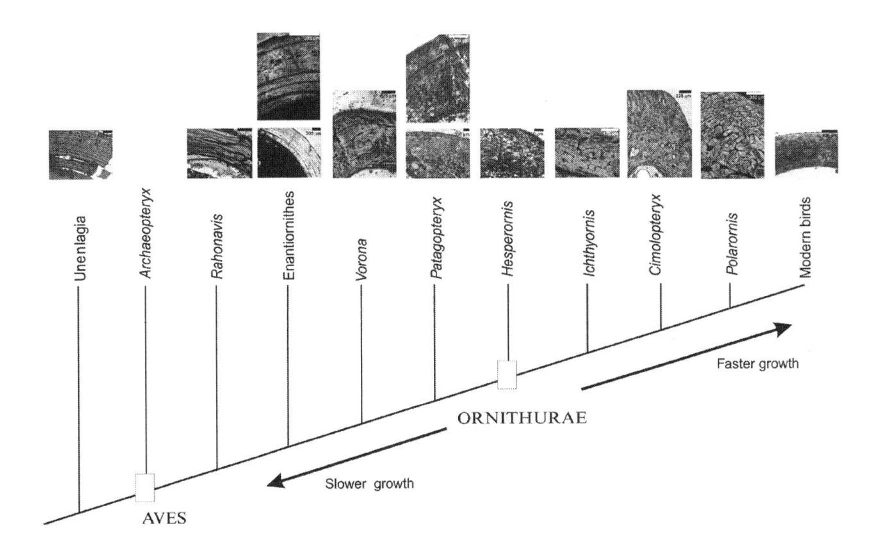

Fig. 6.6. Relationships of Mesozoic birds studied and their bone microstructure. Note the lines of arrested growth in the nonornithurine birds studied: *Rahona, Vorona,* enantiornithines, and *Patagopteryx,* suggesting that growth was periodically interrupted and hence slower than the ornithurine birds *Hesperornis, Ichthyornis, Cimolopteryx,* and *Polarornis.* Notice that the enantiornithine bone is also very poorly vascularized and that enlarged osteocyte lacunae occur.

which indicates a rapid sustained rate of bone formation and, unlike *Patagopteryx* and the two enantiornithine birds studied earlier, showed no trace of arrested growth (Fig. 6.6). The type of bone tissue and overall growth was therefore similar to that of modern birds.

As early as 1926, Heilmann recognized the ecomorphological similarity of *Hesperornis* to loons, and there is consensus that Hesperonithiformes were flightless diving forms that used their laterally compressed feet for propulsion during swimming (in the same way as modern loons and grebes do; Martin 1983). *Hesperornis* bone microstructure appears to reflect its aquatic lifestyle (see chapter 3). The compacta of *Hesperornis* is fairly thick (measuring about 5.2 mm), as compared with the flying birds, and corresponds well with the bone structure of the penguins (e.g., the emperor penguin, which tends to have thick dense compact bone walls of about 5.4 mm), which represents about 33% of the cross-sectional diameter of the bone, and the Humboldt penguin, which has a relative bone wall thickness of 31% (Chinsamy, Martin, and Dodson 1998).

Thickening of the compacta was also evident in the femur of a Cretaceous loon from the Late Cretaceous, Lopez de Bartodano Formation of Seymour Island, Antarctica

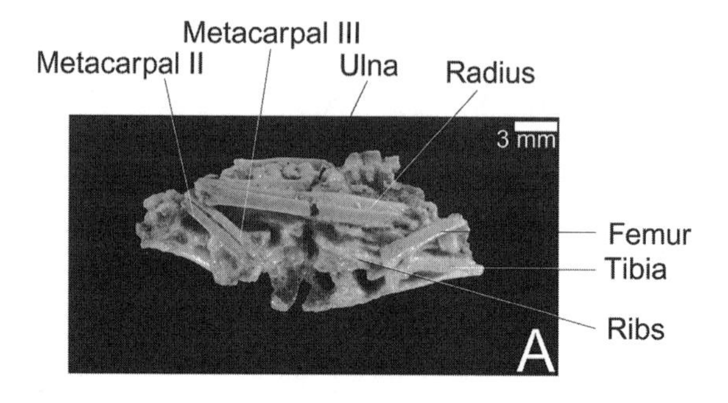

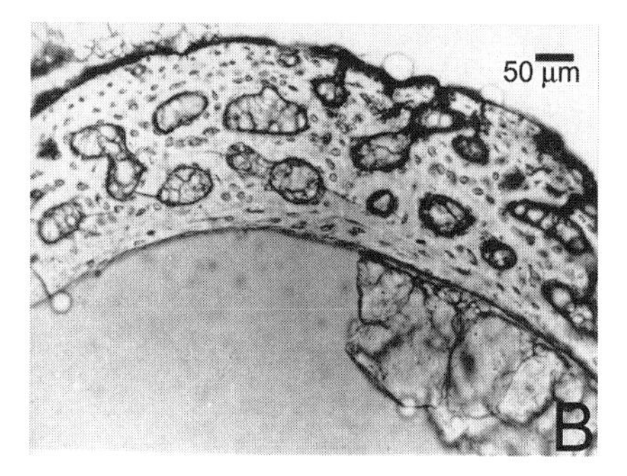

Fig. 6.7. *A.* Left lateral view of the exquisite *Gobipteryx minuta* embryo (ZPAL-MgR-I/34) from the red beds of Khermeen Tsav in the Gobi Desert, Mongolia. *B.* Bone microstructure of the tibia. Courtesy of Andrzej Elzanowski.

(named *Polarornis* by Sankar Chatterjee), which had an RBT value of 37% (Chinsamy, Martin, and Dodson 1998). The thick compacta of this Cretaceous loon consists of numerous primary and secondary osteons, which are embedded in a fibrolamellar bone tissue. Secondary osteons tend to be located in the midcortical-perimedullary region of the compacta, and several large erosion cavities also occur near the medullary cavity, which is lined by a narrow band of circumferentially lamellated bone tissue. Here, too, like the other ornithurine birds studied, no LAGs interrupted the rapid rate of bone deposition (Fig. 6.6).

It is apparent that all the aquatic birds studied showed adaptations in the compact bone to facilitate their particular lifestyle. It is possible that the differences in the actual bone wall thickness might reflect adaptations in their ability to dive to different depths (see chapter 3).

Ichthyornis, the volant bird with its distinctively extremely well developed keeled sternum that facilitated powerful flight, is widely distributed in marine Cretaceous deposits of North America (Martin 1983; Feduccia 1996). Its long jaws with recurved teeth were probably used to capture fish from the water. In terms of phylogenetic relationships, *Ichthyornis* is considered to be an ornithurine bird, though distinct from the neornithines (i.e., modern birds) (Fig. 6.2). *Ichthyornis* was contemporaneous with *Hesperornis* but had very different adaptations in its skeleton related to flight, such as its thin-walled bone structure. The bone microstructure of a humeral fragment, recovered from the Niobrara Chalk of western Kansas, showed that it consisted of fibrolamellar bone tissue without any interruptions (i.e., annuli or LAGs). The channels in the bone in which the blood vessels are located tended to be mainly longitudinally arranged, with some radial and circumferential anastomoses evident. Large erosion spaces are present near the medullary margin.

Another Cretaceous bird that I studied is *Cimolopteryx,* the ornithurine transitional shorebird (Feduccia 1996) from the Hell Creek Formation in Montana. Like the other ornithurine birds, *Cimolopteryx* bone tissue was richly vascularized fibrolamellar bone tissue, and it also did not have any lines of arrested growth present (Chinsamy, Martin, and Dodson 1998; Fig. 6.6). A number of longitudinally oriented primary osteons occur, and several of these are enlarged because of secondary reconstruction.

From the analyses of four Cretaceous ornithurine birds, it is evident that the bone microstructure suggests a rapid rate of bone deposition, without evidence of periodically interrupted growth. It is interesting that none of the ornithurine birds studied show the OCL (typically formed in extant birds when they reach adult body size), which suggests that appositional diametric growth had not yet been completed in these birds and that they were not yet adult in size. However, studies on *Rhea, Casuarius* (Amprino and Godina 1947), and *Struthio* (Chinsamy 1995a) seem to suggest that the OCL can be poorly developed or not at all present in some extant bird taxa. This needs to be further investigated.

Growth Strategies of Basal Birds

The analyses of the bone microstructure of eight nonornithurine bird taxa (*Patagopteryx,* two Argentinean enantiornithines, *Vorona, Rahonavis,* a Malagasy enantiornithine, *Gobipteryx,* and *Confuciusornis*) studied appear to have had periodic arrests in the rate of bone formation, which implies that their growth was episodic. However, the bone microstructure of four Cretaceous ornithurine birds studied to date (*Hesperornis, Ichythyornis, Cimolopteryx,* and *Polarornis*), suggests a rapid rate of bone deposition without any arrests or slowing down in the overall rate of growth. Considering these characteristics within a phylogenetic context suggests that a faster growth without any interruptions characterized the evolution of the Ornithurae. (Chinsamy, Martin, and Dodson 1998; Chinsamy 2002; Fig. 6.6).

Although we are aware that bone microstructure first and foremost provides indications of the relative rate at which bone formed, in extinct animals, on the basis of the different growth strategies evident among the Ornithurae and the nonornithurines, my colleagues and I have suggested that perhaps physiological differences may underlie the difference in growth strategies (e.g., Chinsamy et al. 1994, 1995; Chinsamy, Martin, and Dodson 1998; Chinsamy 2002). Ontogenetic characteristics in bone tissue structure have been documented in the enantiornithines with typical embryonic bone tissue occurring in the embryonic *Gobipteryx,* which suggests that it formed at a rapid rate, while two adult enantiornithines described earlier by Chinsamy and colleagues (1994, 1995) consisted essentially of poorly vascularized parallel-fibered bone. Elzanowski and I have suggested that it is possible that all the rapidly formed bone occured during early ontogeny and was resorbed during growth and remodeling of the skeletal element, which explains why in the adult enantiornithines only the poorly vascularized, parallel-fibered bone tissue is preserved (Chinsamy and Elzanowski 2001). This reasoning is supported by the fact that in the Malagasy enantiornithine, as well as in *Vorona* and *Rahonavis,* an early, more rapidly formed bone tissue comprising fibrolamellar bone is followed by a more slowly formed lamellar type of bone tissue. In *Rahonavis,* the outer layer of more slowly formed lamellar bone makes up about 44% of the compacta, which is comparably much thicker than the circumferentially formed lamellar layer of bone in modern birds. This two-phase growth pattern (i.e., rapid early growth, followed by slower growth) characterizes the growth pattern of early Cretaceous nonornithurine birds (Chinsamy and Elzanowski 2001). We have proposed that (particularly in the enantiornithines) the reduction of the rapid early phase of growth could have been correlated with superprecociality (i.e., independent young, able to fly after hatching), which was previously proposed for *Gobipteryx* on the basis of the high degree of ossification of the wing skeleton and pectoral girdle (Elzanowski 1981). Luis Chiappe and Gareth Dyke (2002) have recently reported that newly discovered enantiornithine juveniles with fledged wings, suggesting the early onset of aerodynamic capability, may indeed support the hypothesis of superprecociality among the enantiornithes. On this basis, Chinsamy and Elzanowski (2001) have suggested that in the enantiornithines the early slowing down of postnatal growth may have been linked to the early onset of flight. Padian and colleagues (2001) have hypothesized that the decrease in the rapid phase of growth in basal birds is linked to the trend in decrease in body size from theropods to birds. More recently, Starck and Chinsamy (2002) have proposed that the change in growth strategy from several years in nonornithurine birds to less than a single year in the ornithurines is linked to strong selection pressure for rapid growth to adult size.

Differences in Growth Rates among the Nonornithurines

Of the eight nonornithurine birds studied, distinct differences in the rate of bone formation have occurred. Although they all form bone in a periodic manner, the bone tissues formed during the period of fast growth appears to be quite variable. In *Patagopteryx,* the annulus suggests a slowing down in the rate of bone formation, which is then followed by a LAG that indicates a pause in the rate of bone deposition and hence growth. The other nonornithurine birds appear to have abrupt pauses in the rate of bone deposition because annuli are not visible before or after the deposition of the LAGs. Furthermore, in *Patagopteryx,* the bone preceding and following the LAG and annulus is fibrolamellar bone tissue. The enantiornithine bones are distinctive in having almost the entire bone wall (except for the endosteally formed circumferential lamellated bone around the medullary cavity) consisting of parallel-fibered bone tissue. The presence of the tide line marks the furthest extent of removal of primary periosteal bone tissue, and given our findings of fibrolamellar bone in *Gobipteryx,* it is possible that this earlier formed bone was removed during medullary expansion during growth. The fact that none of the earlier bone remains, and the fact that most of the bone wall consists of the more slowly formed parallel-fibered bone, suggests that perhaps in the enantiornithines, growth slowed down earlier, perhaps because of energy directed toward a more superprecocial lifestyle. In general, however, the adult enantiornithines appear to be fairly slow growing because LAGs also interrupt the deposition of parallel-fibered bone.

LAGs in Bird Bones

It is well established that LAGs are indicative of pauses or arrests in the rate of bone deposition. The presence or absence of LAGs appears to be a distinctive characteristic in the evolution of birds and directly implies a different growth strategy between basal nonornithurine birds in which bone deposition is punctuated by LAGs and the more derived ornithurine forms where LAGs do not interrupt the rapid growth phase (though they may occur in the peripheral OCL of the bone wall). Some researchers feel that that too much emphasis is placed on the LAGs, and that these are simply plesiomorphic characteristics without any functional significance (Horner et al. 1999). I am, however, against the selective use of data and feel that when interpreting growth strategies of dinosaurs, all aspects of the bone must be considered.

In an attempt to understand the circumstances under which LAGs form, Matthias Starck and I decided to undertake a study of the extant bird *Cortunix japonica* (Japanese quail) to ascertain whether nutritional stress may cause a LAG to form. We subjected a group of 10-day-old hatchlings to a restricted diet, while a control group was allowed to feed ad lib. and monitored their growth for the next 10 days. Our results showed that, although the growth of the experimental birds was effectively stunted and that the bone microstructure was affected by the nutritionally deprived diet in that less bone was deposited and the channels in the bones were much smaller (Starck and Chinsamy 2002), LAGs did not form in the bone.

In my study of subadult elephants that had survived a severe period of drought, I also found no sign of growth rings in their bone tissues, and x-ray analyses also showed no trace of Harris lines in the distal/proximal ends of their bones. In our study of wild birds that have suffered some form of trauma (e.g., injury, disease), Allison Turmakin-Deratzian and I (2003, unpublished data) also did not locate any LAGs in their bones. It therefore seems that some other, perhaps intrinsic, factor is responsible for the arrest of bone deposition in bones because it appears not to be manifested as commonly in bones of extant birds as compared with many nonavian dinosaurs and all nonornithurine birds studied so far. This fact has led Matthias Starck and me (2002) to propose that the nonornithurine birds retain the pleisiomorphic flexible growth patterns of their theropod ancestors, while the ornithurine birds lack such developmental plasticity, perhaps in response to strong selection pressure for rapid growth in less than a year. This probably explains why LAGs can occur in birds, such as *Diatryma* and *Diornis,* with less selection pressure for rapid growth.

Our hypothesis that flexible growth can be retained in birds with extended periods of growth is clearly supported by a recent study of growth in *Aptenodytes patagonicus* (King penguin) chicks. At the 2004 International Comparative Vertebrate Morphology meeting in Boca Raton, Florida, Jacques Castanet presented a paper wherein he showed that when 3-month-old King penguin chicks experience their first austral winter, bone deposition is arrested and a LAG is recorded in their bones. After the 4.5 months of fasting, the parents feed them intensively and bone deposition resumes at rates of up to 171 microns per day forming radial fibrolamellar bone tissue. (This is the highest recorded rate of bone deposition in any vertebrate; see de Margerie et al. 2004.)

Conclusions

Until recently most of what we knew about birds stemmed from anatomical descriptions and functional studies pertaining to flight. Our work on the bone microstructure of basal birds has contributed significantly to understanding the growth and developmental patterns of early birds (Chinsamy et al. 1994, 1995; Chinsamy, Martin, and Dodson 1998; Chinsamy and Elzanowski 2001; Starck and Chinsamy 2002). By considering the bone microstructure of basal birds in a phylogenetic and ontogenetic context, our work has shown that basal birds grew at rates that were much slower than that of modern birds, requiring several years to reach a mature body size (Chinsamy et al. 1994; Chinsamy, Martin, and Dodson 1998; Chinsamy and Elzanowski 2001; Chinsamy 2002; Starck and Chinsamy 2002). It is possible that the reduction in the amount of rapidly formed bone may be linked to a reduction in the overall size of basal birds as compared with their nonavian theropod ancestors (Padian et al. 2001), although it is also possible that (particularly in the enantiornithines) it may also be linked to the onset of precocial flight as suggested by Chinsamy and Elzanowski (2001).

In trying to understand the differences in bone tissues of basal birds, Matthias Starck and I (2002) have suggested that the change from an interrupted to an uninterrupted growth pattern from the nonornithurine to the ornithurine birds may be linked to the loss of developmental plasticity in the more derived birds. This hypothesis is supported by the fact that the nonavian dinosaurian ancestors of birds had the ability for flexible growth patterns (bone with LAGs vs. uninterrupted bone), but perhaps with the selection pressures for fast growth, the more derived ornithurines lose this pleisiomorphic characteristic of flexible growth. This is supported by the fact that large, ground-dwelling birds with reduced selection pressures for rapid growth (such as *Diatryma*) have retained the ability for flexible growth patterns and show bone microstructure with periodic interruptions. Indeed King penguins also retain a flexible growth strategy because of their unique growth strategy.

Given that our knowledge of Mesozoic birds has more than trebled over the past 20 years, and the phenomenal rate at which new fossils are being discovered, it is only reasonable to expect that our understanding of early bird radiation and biology will be significantly enhanced over the next few years.

CHAPTER SEVEN

Dinosaur Physiology

One of the most intriguing questions about dinosaurs is whether they were endotherms (warm-blooded) or ectotherms (cold-blooded). This ongoing, decades-long debate still has not reached a consensus. Studying the physiology of living animals poses many problems. It should therefore come as no surprise that questions about the physiology of animals dead for 65 million years or more is rife with controversy and speculation. Thus, the biggest challenge dinosaur paleobiologists face is that the physiology of dinosaurs cannot be readily observed or assessed.

Because most of what we know about dinosaurs is derived from the preserved remains of their skeletons with virtually minimal information about soft tissue structures, assessing their physiology is no easy task. As such, various lines of evidence have been invoked to understand the physiology of dinosaurs. However, progress in this area has been impeded by the lack of objectivity of researchers. As dinosaur paleontologist Jim Farlow of Indiana University–Purdue University in Fort Wayne says, they behave more like "politicians and lawyers" compelled to "defend" their hypotheses (Farlow 1990). Compounding this is the fact that several researchers have tended to presume that single characteristics (e.g., such as posture and brain size) are diagnostic of a whole physiological complex.

It is worth bearing in mind that ectothermy is a primitive condition for amniotes and, indeed, all vertebrates, while endothermy is a derived condition in mammals and birds and evolved independently in these lineages (Chinsamy and Hillenius 2004). Given the phylogenetic relationship of nonavian dinosaurs to extant birds, exactly where and when endothermy evolved is indeed compelling. In the debates around this, the two main arguments are that:

1. Endothermy evolved among the nonavian dinosaurs and was inherited by birds (Aves);
2. Endothermy occurred late in the evolution of the Dinosauria and coincided with the origin of the Ornithurae.

The terms *warm-blooded* and *cold-blooded* have permeated the popular media with much vigor but are rather simplistic and, frankly, scientifically inadequate. In many instances, there is an erroneous equation of thermoregulation (i.e., the regulation of body temperature) and aerobic capacity (i.e., the capacity for sustained oxidative metabolism). Furthermore, *warm-bloodedness* is often uncritically equated with homeothermy and endothermy, as well as with high levels of metabolic rates, while ectothermy is often used synonymously for *poikilothermy* and *cold-bloodedness*. This oversimplification fails to recognize the complexities of the thermal biology of endotherms and ectotherms.

Endothermic animals (endotherms) have high rates of internal heat production that enables them to maintain homeothermy (i.e., a relatively constant body temperature), even though the environmental temperature is variable. Ectothermic animals (ectotherms) are incapable of maintaining homeothermy in changing ambient thermal conditions and are referred to as *poikilothermic*. Ectotherms depend on heat in the environment to maintain their body temperature. This, however, does not mean that they cannot attain homeothermy. Under favorable conditions, they are able to maintain a significant thermal gradient between their body temperature and that of the environment. Many ectotherms, such as lizards, alter behavior to thermoregulate effectively. Large ectotherms have a high thermal inertia because of their small surface area to volume ratio, which reduces heat loss, and can be effectively homeothermic. Physiologist Frank Paladino and his colleagues (1997) speculated that thermal inertia would increase with increasing body size and deduced that large dinosaurs could easily have been homeothermic simply because of their gigantic body sizes. The term *gigantothermy* is used for this type of thermal biology. Extensive studies on extant vertebrates suggest that the thermal profiles of endotherms and ectotherms can converge under certain circumstances.

A fundamental difference between ectotherms and endotherms is their inherent aerobic capacities (i.e., their capacity for sustained oxidative metabolism). Endotherms generally have metabolic rates that are about ten to fifteen times greater than that of ectotherms of the same body mass and temperature. To support such rates of aerobiosis, and assimilation of oxygen to tissues, endotherms have many specialized adaptations, such as specialized lungs with increased pulmonary capacities and ventilation rates, fully separated pulmonary and systemic circulations, increased blood volumes and blood oxygen-carrying capacity, as well as increased mitochondrial density and enzymatic activities. Thus, a tachymetabolic endotherm is capable of having a sustained high rate of metabolism irrespective of environmental constraints, while an animal that is bradymetabolic has a low level of metabolism. Another term not as commonly used but one that is equally important is *heterothermic*, which implies a variable body temperature.

Being endothermic or ectothermic has consequences regarding virtually every aspect of an animal's overall biology. Tachymetabolic endothermy imparts a high rate of metabolism to an animal, which is manifested by, for example, rapid growth rates, high aerobic capacity, and sustained activity. Unfortunately, the fossil record does not preserve any of the structural and functional modifications that are required to support high levels of aerobic metabolism. Consequently, paleobiologists are tasked with deciphering by other means the thermal biology and aerobic capacity of dinosaurs, which has led to much speculation, controversy, and heated debates about the physiology of dinosaurs.

In the early part of the nineteenth century, dinosaurs were generally thought to be "oversized" reptiles and as such were often portrayed as sluggish, dim-witted, cold-blooded creatures. By the latter part of the nineteenth century, Thomas Huxley recognized similarities between the earliest bird, *Archaeopteryx,* and small theropod dinosaurs. However, this idea of a more advanced physiology for dinosaurs did not last, and people continued to perceive dinosaurs as giant reptiles. In the 1970s, John Ostrom revived Huxley's proposal that birds and dinosaurs were related, and Robert (Bob) Bakker took up these ideas and began promoting his concept of dinosaurs as dynamic endotherms. Among the various methods that he used to make this deduction (such as predator-prey ratios, posture, and locomotory ability), he also used characteristics in dinosaur bone microstructure. Because of his charismatic style and exceptional ability to convey his ideas, Bob Bakker successfully managed to further stimulate a "renaissance" in the perception of dinosaurs. Today, the media and general public generally perceive and portray dinosaurs as highly active, endothermic animals. It is perhaps important to mention that this notion of "active" dinosaurs is based on the erroneous assumption that only endothermy can sustain the dynamic attributes of dinosaurs. Recent research on ectotherms has shown that their thermal biology and aerobic capacities are a whole lot more complex than previously supposed and that they need not necessarily be portrayed as sluggish and inactive.

In the quest to understand the physiology and more specifically the thermal biology of dinosaurs, several aspects of dinosaur anatomy, ecology, and phylogeny have been used. Studies of dinosaur locomotory apparatus, predator/prey ratios, brain size, biomechanical energetics, social behavior, bird origins, and bone histology, among others, have all been used to assess the physiology of dinosaurs. More recently, oxygen isotope analysis and turbinate bones have also been invoked. Even after more than two decades of intensive research into the question of dinosaur physiology, there is no panacea. In recent years, several reviews of dinosaur physiology have been published, for example, in *The Dinosauria* (Farlow 1990), in the *Annual Review of Ecology and Systematics* (Farlow et al. 1995), and more recently in the more popular "*The Complete Dinosaur*" (Reid 1997). Jaap Hillenius and I have reviewed the topic in the latest edition of *The Dinosauria* (Chinsamy and Hillenius 2004) and yet another review appears in *Physiological and Biochemical Zoology* (Hillenius and Ruben 2004). As such, it seems unnecessary for me to present yet another review of all the various arguments here, and I will therefore mainly focus on assessments made from bone microstructure and only briefly touch on more recent work on dinosaur physiology such as nasal turbinates, lung structure and ventilation rates, and oxygen isotopes.

Bone Microstructure

Several early studies on the microscopic study of dinosaur bone, notably by researchers such as Adolf Seitz (1907), Walter Gross (1934), Donald Enlow and Sidney Brown (1957), and John Currey (1960) recognized the high levels of vascularization, and the predominant fibrolamellar bone and occurrence of dense Haversian bone in dinosaur bones (Fig. 3.8C; Plates 5B, 7, 8) as strikingly similar to those of large mammals. The first tentative links between bone microstructure and thermal biology was made by Enlow and Brown (1958) and later John Currey (1962) also raised this possibility. In more recent times however, Armand de Ricqlès (e.g., 1974, 1976, 1980) can be singled out as having brought to the fore the similarity of dinosaur bone to mammalian bone, and for advocating the use of bone histology for interpreting thermal biology of extinct animals. In addition to the presence of fibrolamellar bone with extensive vascularization and Haversian bone, de Ricqlès (1976, 1980) considered the absence of cyclical growth marks in dinosaur bone as important for deducing physiological attributes. He saw these histological characteristics of dinosaur bone as indicative of high rates of sustained growth, intensive bone/body fluid exchange, high metabolic rates, and suggestive of endothermy. De Ricqlès (1983, 1992) later reiterated his views and proposed that dinosaurs may not have been endothermic like modern birds and mammals but that they may have had an intermediate type of physiology.

Since the early 1970s, Bob Bakker argued that dinosaurs had a typical, modern avian-type physiology with homeothermy attained through endothermy, and he later used characteristics of dinosaur bone microstructure in support of his hypothesis (Bakker 1986). Although they differed in their overall deductions, both Bakker and de Ricqlès advocated the use of bone microstructure to infer thermal physiology of dinosaurs.

Robin Reid began writing about dinosaur bone tissues in the early 1980s, and he was the first person to point out that contrary to de Ricqlès (pre-1983), not all dinosaurs grew in a rapid, sustained manner like extant endotherms. Through his cogent and thorough analyses of dinosaurian bone tissues, Reid expounded on the limitations of using bone to make physiological deductions in various publications. Reid highlighted the important fact that there is no distinctive characteristic of bone in endotherms or ectotherms and cautioned against using bone tissues as indicators of thermal physiology and metabolic rates (e.g., Reid 1984a, 1984b, 1987). However, Reid recognized that within these limitations, reasonable inferences can be made from dinosaurian bone tissues. For Reid, the abundant presence of fibrolamellar bone implies a capacity for rapid growth, while zonal bone implied that some dinosaurs grew in a cyclical manner and were apparently affected by ambient conditions. Reid (1997) proposed that dinosaurs were "failed endotherms," that is, they had made some progress toward endothermy over the typical ectothermic levels (through having double-pump circulation and aerobic activity) but that this progress was not at the same level as extant endotherms because dinosaurs appeared to rely on bulk to maintain homeothermy.

When I began working on dinosaur bone microstructure in 1986, Reid had already published his findings of zonal bone in a sauropod dinosaur and, in so doing, raised a "red flag" about the problems of interpreting physiology from bone tissue structure. As such, I have been cautious in making physiological deductions from the microscopic structure of dinosaur bones. I have always maintained that interpretations of what dinosaur bone microstructure means in terms of physiology is debatable, but the actual microstructure of dinosaur bone provides undisputed evidence of how the particular bone actually grew (e.g., Chinsamy 1994, 1997; Farlow et al. 1995).

I agree with de Ricqlès and Reid that fibrolamellar bone, as suggested by Amprino (1947), implies that bone was deposited at a rapid rate, although contra de Ricqlès, I endorse Reid's view that high metabolic rates are not required for fibrolamellar bone to be formed and that such tissues do not necessarily imply endothermy. I also concur with Reid that the presence of zonal bone tissue (periodic lines of arrested growth [LAGs]/annuli) in dinosaur bones is significant and indicates a fluctuating rate of growth, which contrasts with Horner and colleagues (1999) who see LAGs as simply plesiomorphic characteristics without functional significance. Much to my chagrin, I am sometimes cited completely out of context, misrepresented, and even misquoted (Padian and Horner 2004, p. 1879) as making direct physiological deductions from bone. Sure, like *all* researchers in this field, I have speculated about what particular bone tissue types might mean in terms of physiology, but I have always emphasized this as an indirect inference (e.g., Chinsamy 1994; Chinsamy et al. 1994). This is perhaps a good place to reiterate my view that bone microstructure directly indicates how bone formed during growth and provides indications of the factors that may have affected its growth, such as seasonality, ontogenetic age, and lifestyle adaptations, while deductions made about physiology are speculative and must be recognized as such (e.g., Chinsamy 1994, 1997; Farlow et al. 1995). This view is contra Padian (1997, p. 554) who asserts that the histology of bones "is the most direct line of preserved paleontological evidence about cellular physiology because bone cells and tissues are direct records of metabolic activity."

Characteristics such as the extent of Haversian bone formation, fibrolamellar bone tissue, and high vascularity of dinosaur bone microstructure (Plates 7, 8) were essentially used by Armand de Ricqlès and Bob Bakker to make physiological deductions, and each of these characteristics, as well as zonal bone, are considered below to understand the interpretations made from them, as well as what our current views are about them.

Fibrolamellar Bone and High Vascularity

Fibrolamellar bone and its associated high vascularity, typically occurs in the bones of mammals and birds (see Plates 2, 7). The prevalence of fibrolamellar bone as primary compact bone among various dinosaurs led de Ricqlès to suggest that they inferred rapid growth and high metabolic rates; a view currently supported by both Kevin Padian and Jack Horner. Bakker also used the frequent occurrence of fibrolamellar tissue in dinosaur bones to infer an endothermic physiology for dinosaurs.

Such deductions failed to consider the wide range of animals that are known to produce fibrolamellar bone: juvenile crocodiles farmed on commercial crocodile farms, where optimal temperatures for growth are provided, are able to form fibrolamellar bone in the zonal regions of their compact bone. However, fibrolamellar bone has been described in the bones of wild animals, for example, by Reid, in the bones of feral *Alligator mississippiensis* and *Crocodylus porosus,* as well as in a Late Cretaceous turtle and in a Pleistocene gavial (Fig. 7.1). And in their classic studies of vertebrate bone, Donald

Enlow and Sidney Brown (1957) found fibrolamellar bone in pelycosaurs and captorhinids, and Enlow (1969) describes and illustrates fibrolamellar bone in the zones of a midshaft section of a 15-year-old Nile crocodile (*Crocodylus niloticus*). Armand de Ricqlès (1972) has documented fibrolamellar bone in a variety of nonmammalian synapsids (e.g., deinocephalians, dicynodonts, gorponopsians, therecephalians, and cynodonts), and Walter Gross (1934) also showed the presence of such tissue in dicynodontians. In a study of humeri of several genera of dicynodonts (*Aulacephalodon, Cistecephalus, Dicynodon, Endothiodon, Lystrosaurus, Kannemeyeria, Oudenodon*) from various stratigraphic horizons of the Beaufort Group in South Africa, Bruce Rubidge and I documented the predominance of fibrolamellar bone in the zonal regions of their stratified compacta (Chinsamy and Rubidge 1993). Only in the small, burrowing form of *Diictodon* did we find uninterrupted fibrolamellar bone (Fig. 7.2). However, more recent work incorporating several skeletal elements and several growth stages of *Diictodon* has found that interrupted bone occurs in individuals that are greater than 70% of adult size (Ray and Chinsamy 2004). These findings suggest that for most of their lives *Diictodon* grew relatively fast and that only much later in their ontogeny did they grow at slower rates. In addition, my recently graduated PhD student Jennifer Botha has reported interrupted and uninterrupted fibrolamellar bone tissues among six genera of nonmammalian cynodonts studied (Botha 2001). In a recent publication, we reported that contemporaneous cynodonts, *Diademodon* and *Cynognathus,* which are often also found together in fossil assemblages, experienced starkly different growth strategies (Botha and Chinsamy 2000). Both taxa formed predominantly fibrolamellar bone, but *Diademodon* experienced periodic pauses in its growth, while *Cynognathus* grew in an uninterrupted, sustained manner (Fig. 7.2). Another interesting finding is that the evolution of the Cynodontia seems to be characterized by the increasingly rapid rates of growth; for example, the basal *Procynosuchus* had the slowest growth, while *Tritylodon,* the most derived taxon included in this study, showed uninterrupted fibrolamellar bone tissues, which suggests rapid sustained growth (Botha 2001; Ray 2004). Fibrolamellar bone has also been documented in several skeletal elements of the Late Permian gorgonopsian, *Scylacops* (Ray et al. 2004).

Thus, it is evident that fibrolamellar bone is not restricted to nonavian dinosaurs, mammals, and birds. Its widespread occurrence among the Reptilia, and among animals that are generally considered to be ectotherms, such as the Permian dicynodonts and gorgonopsians suggest that the ability to form this bone tissue is evidently not linked to high metabolic rates. Given the wide distribution of fibrolamellar bone, the claim of a correlation between fibrolamellar bone and high metabolic rates and endothermy appears unsubstantiated (Chinsamy 1994; Chinsamy and Hillenius 2004).

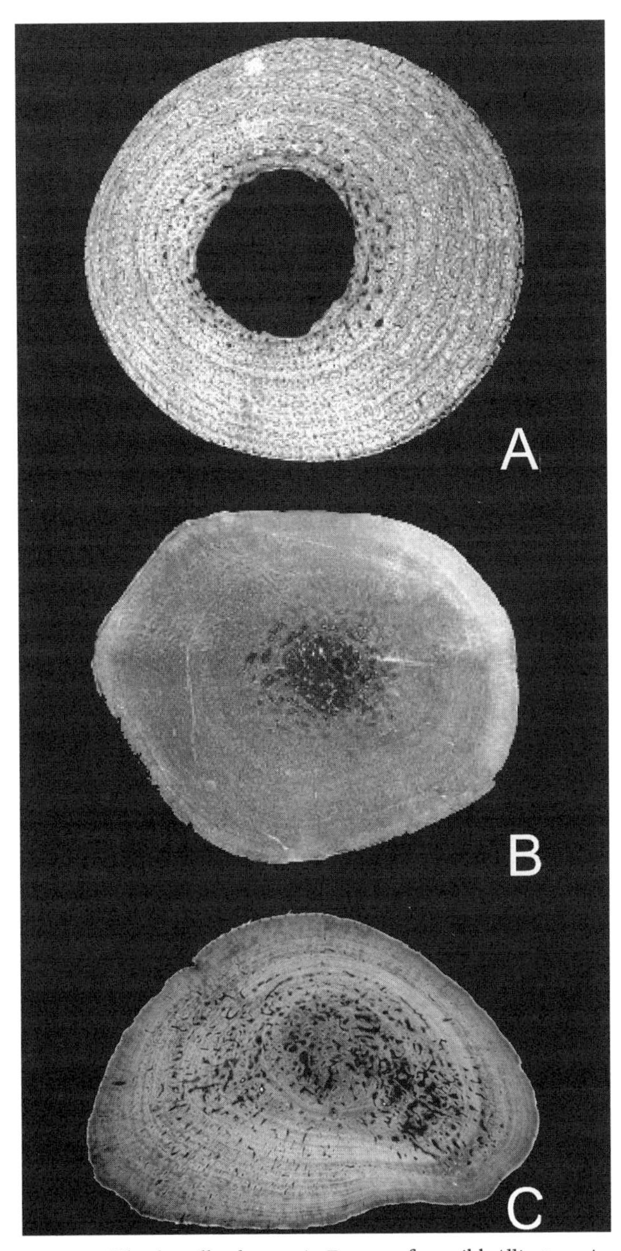

Fig. 7.1. Fibrolamellar bone. *A*. Femur of a wild *Alligator mississipiensis* from North Carolina. Cross-sectional diameter measures 11 mm. *B*. A femur of a wild crocodile from Queensland Australia (maximum diameter, 20 mm). *C*. A scapula of a freshwater turtle (maximum diameter, 16 mm) from the Late Cretaceous of Montana. Images courtesy of Robin Reid.

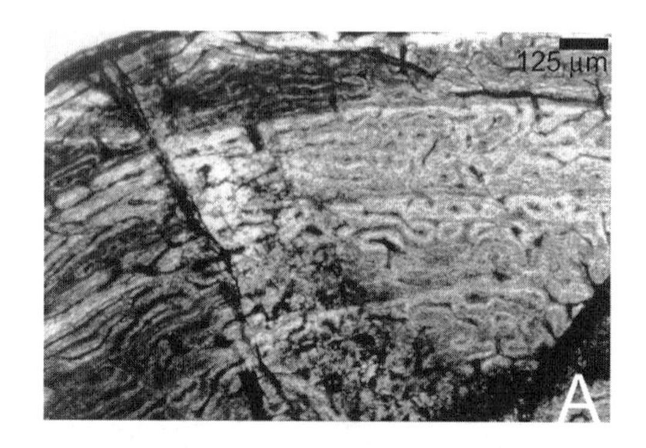

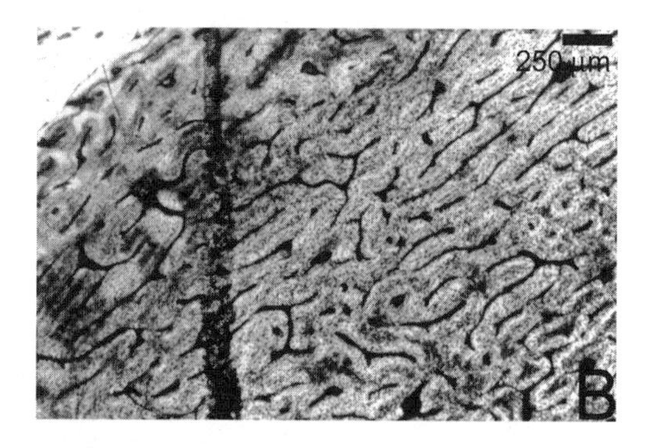

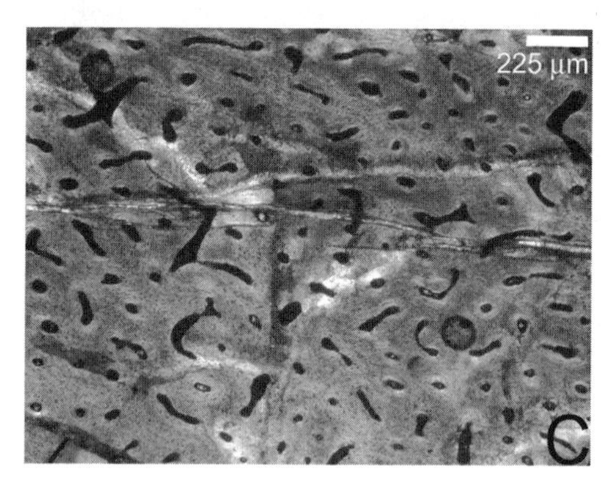

Fig. 7.2. Fibrolamellar bone in nonmammalian therapsids. *A. Diademodon* femur, fibrolamellar bone in the zonal regions of the bone. *B.* Azonal fibrolamellar bone in a *Cynognathus* femur. *C.* Fibrolamellar bone in a *Diictodon* humerus. *A* and *B* courtesy of Jennifer Botha.

Recent work by Kevin Padian, Armand de Ricqlès, and Jack Horner (e.g., 2001) has contended that it is not just the presence of fibrolamellar bone in dinosaurs that is significant but also the extent to which it occurs and the type of vascularization therein. This argument does not consider that the size and shape of the so-called vascular channels of fossil bone have no real bearing on the size and shape of the actual blood vessels therein and also does not consider ontogenetic changes in the orientation of the channels in the bone. For example, hatchling and young birds, even though they are growing very fast, tend to have a predominance of longitudinally orientated vascular canals (e.g., Chinsamy 1995a; Starck and Chinsamy 2002). Only later in ontogeny do these join to form the reticular and circumferential networks that Padian et al. (2001) refer to as characteristic of fast growth rates of birds. Furthermore, even within a single cross section of a single bone, let alone in different bones of the skeleton, the type of fibrolamellar bone and its associated "vascularization" can vary in response to local conditions (e.g., vicinity of trochanters or muscular insertions). I believe that fibrolamellar bone cannot be used to infer endothermy or high metabolic rates, although I agree that its presence is indicative that bone deposition occurred at a rapid rate (i.e., relative to other bone tissues such as lamellar bone tissue that may have formed in the animal). But on the basis of evidence from studies on modern bone depositional rates (e.g., Starck and Chinsamy 2002; de Margerie et al. 2004), I do not agree that similar bone tissues are deposited at similar rates (Horner et al. 2000; Padian et al. 2001).

Haversian Bone

In this context, Haversian bone refers to secondary osteons and more particularly to dense Haversian bone in which interstitial bone between neighboring Haversian systems consists of remnants of older ones (see Fig. 4.2; Plate 5B). Walter Gross in 1934 first recognized secondary osteons in the prosauropod *Plateosaurus* and the sauropod *Brachiosaurus*. However, even much earlier, Adolf Seitz (1907) showed that a tissue now called Haversian bone could be developed in dinosaurs at a level comparable to that of large mammals. The fact that Haversian bone is common among dinosaurs (Fig. 4.2; Plate 8) and is usually extensively developed led to bone microstructure initially becoming embroiled in the controversy about dinosaur thermal biology.

Bob Bakker interpreted the Haversian character of dinosaur bone as implying high levels of calcium and phosphate exchange. In his earlier papers, de Ricqlès (e.g., 1972, 1974) also regarded Haversian bone (and the frequent occurrence of fibrolamellar bone as primary compact bone in dinosaurs) as suggesting endothermy, while other researchers took it as evidence of mass homeothermy. Yet others, such as Fuzz Crompton

(1989) of Harvard University, considered Haversian bone to be a consequence of mechanical strain in the bone.

Today, it is well recognized that Haversian bone is in fact not related to high standard metabolic rates (as implied by endothermy), especially because it is not present in small birds and mammals that have the highest metabolic rates, and it is also not restricted to endotherms. It can be extensively developed in reptiles; for example, Reid (1987) has reported the cortex of a femur of the tortoise *Geochelone triserrata* being composed entirely of dense Haversian bone. To complicate matters even further, Haversian bone has been clearly shown to be associated with various nonthermal factors such as muscular attachments and remodeling and frequently increases with increasing age. Experimental studies of *Alligator mississippiensis* farmed at temperatures between 23.5° and 27.0°C show an increase in the amount of Haversian bone as compared with wild forms of similar sizes. This suggests that inertial homeothermy at the level expected for large dinosaurs could have similar effects.

Zonal Bone

It is well recognized that extant poikilothermic, ectothermic vertebrates typically form bone in a periodic, cyclical manner (e.g., Enlow and Brown 1957; Peabody 1961). Even those reptiles that form fibrolamellar bone only do so in the favorable period of growth (i.e., fibrolamellar bone is deposited in the zonal regions of the compacta and is generally followed by an annulus and/or a line of arrested growth). This pattern of cyclical growth contrasts sharply with that of extant mammals and birds, which tend to deposit fibrolamellar bone continuously without any interruptions in bone deposition. However, Klevezal and Kleinenberg (1969) and Klevezal (1996) documented the occurrence of zonal type of bone tissue in several marine mammals, hibernators, and those that experience bouts of periodic cold (see chapter 3). Isolated reports of zonal bone tissues in a polar bear and in a kangaroo are known (Fig. 3.5C, 3.5D), and a LAG has been reported in a single bone of an elk (Horner et al. 1999). Exactly why bone deposition was interrupted in these mammals is unknown, though physiological/intrinsic stress may be an important factor. However, it is evident that such tissues are more the exception than the rule among extant mammals. It would also be worth testing ontogenetically whether the periodicity of such interruptions in extant mammals is annual as is the case in extant reptiles.

Although John Campbell (1966) had recognized the reptilian "annualate structure" of dinosaur bone, pre-1981 dinosaurs were still generally recognized as having bone compacta made up typically of azonal fibrolamellar bone tissues (e.g., de Ricqlès 1980; see chapter 4). Following Robin Reid's publication of zonal bone in a sauropod pubis, and his subsequent publication in 1990 showing the widespread occurrence of zonal bone among the Dinosauria, it is now well known that dinosaurs were able to form both zonal and azonal bone.

Horner and colleagues (1999) contend that because growth rings and LAGs can occur in extant mammals their occurrence in dinosaurs is irrelevant. However, several studies of ontogenetic series of dinosaurs have consistently shown that numbers of growth rings increase with age and skeletochronology has been applied to obtain age estimates for several dinosaurs (e.g., Chinsamy 1990; Horner et al. 2000a; Erickson et al. 2001). Thus, when genuine growth rings (i.e., not confused with other growth marks such as tide lines and reversal lines) are discerned in dinosaurs, they are episodic and interrupt the rapid rate of bone formation. Such fluctuating, flexible growth patterns characterize growth in extant poikilothermic reptiles. However, this does not imply that dinosaurs grew at the same rates as extant reptiles because rapidly formed fibrolamellar bone is usually present in the zonal regions of the compacta of dinosaurs.

Reid has written several articles in which arguments against the use of bone microstructure to deduce high metabolic rates and endothermy have been substantiated and has pointed out that the presence of growth rings in some dinosaurs implies their inability to grow at a sustained rapid rate. Considering both bone microstructure and hemodynamics, Reid formulated a hypothesis that proposes that dinosaurs were probably intermediate or "failed endotherms" (Reid 1987, 1997). He suggested that dinosaurs took two critical steps toward endothermy (i.e., a double-pump circulation and high aerobic activity), but then used bulk as a means to stabilize their body temperature. He has proposed that dinosaurs may not have been endotherms but could have been subendothermic, "superreptiles," or "supergigantotherms" having a more advanced cardiovascular system than any modern reptile and no modern physiological representative.

As a young graduate student, in my very first publication, I reasoned that the bone microstructure of *Syntarsus,* which showed the typical fibrolamellar bone with well-developed annuli, could be interpreted to imply at least three possible physiologies: that *Syntarsus* was an ectotherm, that *Syntarsus* probably had a thermophysiology intermediate between ectothermy and endothermy, and lastly that *Syntarsus* was a heterothermic endotherm (Chinsamy 1990). As shown in this article, the presence of zonal bone in dinosaurs could be interpreted to fit any of the above scenarios. In my later publications, I became more cautious in relating bone microstructure to physiology and have emphasized that bone microstructure was most useful in indicating growth patterns rather than thermal physiology (e.g., Chinsamy 1994).

Some dinosaurs, in particular, the small to medium-sized *Dryosaurus* that I worked on, the Proctor Lake ornithopods and several Australian Polar hypsilophodontids, which Tom Rich and Pat Vickers-Rich and I studied, exhibited a bone tissue without any interruptions

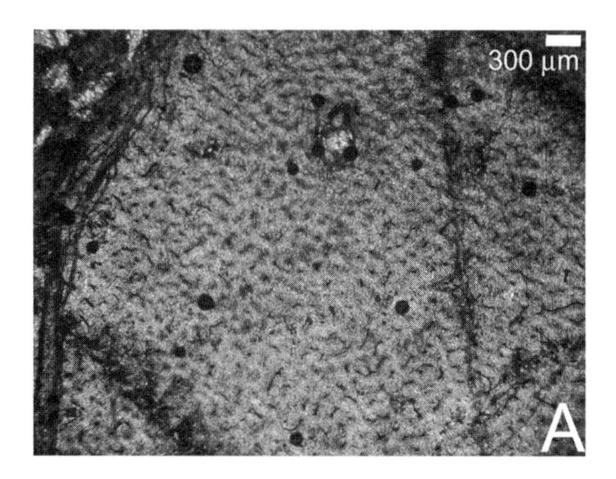

Fig. 7.3. Azonal bone in femora. *A.* The basal theropod, *Herrerasaurus. B. Dryosaurus.*

in bone formation (Fig. 7.3). In addition, Robin Reid has found such azonal bone in other dinosaurs such as *Iguanodon*, *Neovenator*, *Allosaurus*, and *Megalosaurus*. Recently, I had the opportunity of studying the bone microstructure of *Herrerasaurus* (Fig. 7.3), the basal theropod from Argentina, which also showed uninterrupted fibrolamellar bone tissues, and Allison Turmakin-Deratzian has found a similar condition in *Centrosaurus*. Although these are mostly from isolated single bones as opposed to the growth series, they do suggest that azonal bone (i.e., uninterrupted fibrolamellar bone tissue) is quite widespread among the Dinosauria, as de Ricqlès had originally proposed. Considering that dinosaurs were able to form fibrolamellar bone continuously to fairly large sizes points to a distinct difference between them and other reptiles. Contra de Ricqlès (e.g., 1974), Reid has suggested (1996) that the ability to do this was not a result of high metabolic rates but rather high circulatory efficiency. The question remains though, why did some dinosaurs grow in a sustained rapid manner, while others grew in a punctuated, periodic manner?

It is interesting to note that even contemporary dinosaurs, facing the same ambient conditions, apparently grew differently. The following is a case in point: during the Early Cretaceous, the southeastern part of Australia was at about 70°–80° south, well within the Antarctic Circle. Oxygen isotope studies have proposed a temperature range of –2° to ±5°C, although the paleofloral and faunal composition suggest a more moderate 8°–10°C. Nevertheless, although the temperature was more moderate than current ranges, the region would have had long periods of darkness. Surprisingly, a diversity of dinosaurs has been recovered from this high-latitude locality, and Tom Rich, Pat Vickers-Rich, and I (Chinsamy, Rich, and Vickers-Rich 1998) were curious about whether the bone microstructure of these dinosaurs would provide any clues about their growth and lives. Our small but very special sample comprised two hypsilophodontids (one notably that of *Leaellynasaura,* named after the Riches' daughter, Leaellyn), as well as the small theropod dinosaur, *Timimus* (named after the Riches' son, Tim). The hypsilophodontids showed azonal bone, while the contemporaneous *Timimus* showed a compact bone with distinct LAGs (Fig. 7.4). If both the hypsilophodontids and the theropod had zonal bone, we could have reasoned that the periods of arrested growth were the result of the long unfavorable season (as is the case in many hibernating mammals and in polar bears that tend to be less active during the winter periods (see Fig. 3.5D). However, our results showed clearly that the hypsilophodontids grew at sustained uninterrupted rates. They were probably too small to migrate significant distances to escape the cold winters, and absence of growth rings in their bones suggests that they did not hibernate. It is perhaps possible that they reached somatic maturity before the unfavorable season, and therefore no growth rings are recorded. It is interesting that the endocast of the brain of *Leaellynasaura* shows enlarged optic lobes, which suggests that they had enhanced vision, making it quite plausible that they could have remained active during the dark polar winters. Recent work by Andrew Constantine from Monash University in Australia (Constantine et al. 1998) has shown that just 3 m above where one of the hypsilodontids were found sedimentary structures appear to have been formed in seasonally frozen ground. Could the polar hypsilophodontids have been endothermic? Unfortunately, although it is tempting, in the absence of other signals of endothermy (such as high aerobic capacity), we cannot make this deduction. It is however appropriate to deduce that the polar hypsilophodontids had a different growth strategy to *Timimus.*

Evidence is mounting of a "specialized" growth strategy among other hypsilophodontid populations from other localities. For example, the growth series of *Dryosaurus* showed azonal bone (Fig. 5.6), while the contemporary *Kentrosaurus* from

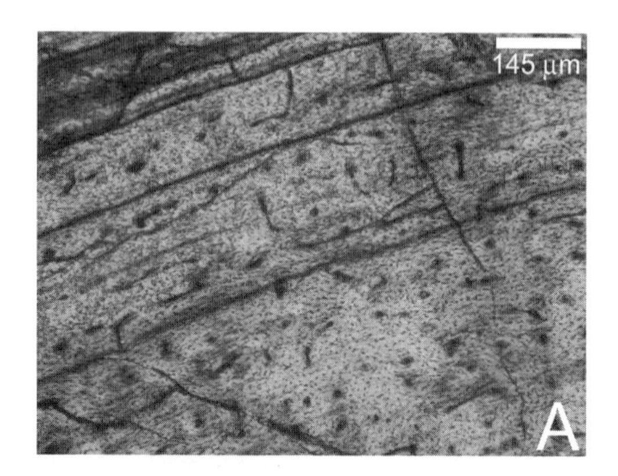

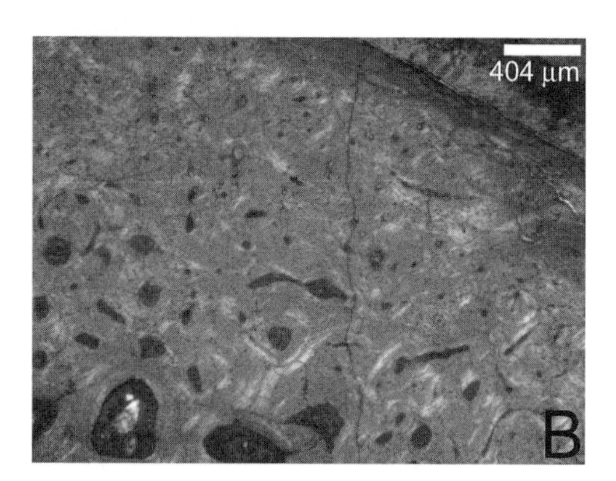

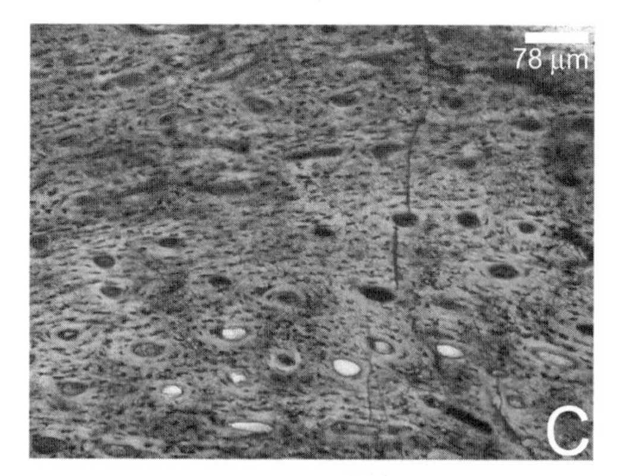

Fig. 7.4. Transverse section of femora of Early Cretaceous polar dinosaurs from Dinosaur Cove, Australia. *A. Timimus,* zonal bone. *B.* Azonal bone in a hypsilophodontid femur. *C.* Azonal bone in *Leaellynasaura.*

the same locality has zonal bone. In 1994, at the Society of Vertebrate Paleontology meeting, Dale Winkler reported that the Proctor Lake ornithopods had azonal bone, while contemporary theropods and hadrosaurs showed periodic interruptions in their bones. Thus, it is apparent that the hypsilophodontids from divergent localities at different latitudes show differences in their growth compared with other contemporary dinosaurs. Physiological parameters may underlie such differences in growth biology, but the nature of this relationship is uncertain since there is no causal connection between bone tissue and basal metabolic rate (e.g., Chinsamy and Hillenius 2004).

In the growth series of *Dryosaurus* that I studied, and in the Proctor Lake forms that Dale Winkler studied, we documented that throughout ontogeny they were able to grow at a sustained rapid rate. An important lesson I learned from Robin Reid over the years is that one should not interpret beyond the evidence presented. Therefore, considering all the factors presented in the bone microstructure of *Dryosaurus,* not just the presence or absence of LAGs, I believe that the rapid rate of growth (indicated by uninterrupted fibrolamellar bone) cannot simply be attributed to endothermy.

The variable occurrence of zonal and azonal bone among the Dinosauria seems to reinforce the hypothesis that rather than physiological status, bone microstructure more accurately reflects the plastic nature of bone and its response to environmental factors (Starck and Chinsamy 2002; Chinsamy and Hillenius 2004).

Growth Rings in Basal Birds

In chapter 6, we saw that the bone microstructure of all eight basal birds studied histologically differs from that of modern birds by having a bone deposition pattern that was periodically interrupted by LAGs, whereas in modern birds, such lines tend to be restricted to the outer circumferential layer of bone and do not interrupt the usual rapid growth of the young birds (Fig. 6.6). Thus, all nonornithurine birds showed periodically interrupted growth as compared with ornithurine forms; even compared with Cretaceous ornithurine forms such as *Hesperornis, Polarornis, Cimolopteryx,* and *Ichthyornis* (Chinsamy et al. 1994, 2002; Chinsamy, Dodson, and Martin 1998; Fig. 6.6). Although my co-workers and I recognize that bone microstructure first and foremost informs about growth patterns, the stark differences in growth between ornithurine and nonornithurine birds seemed to suggest some physiological difference between them, and we proposed that perhaps nonornithurine forms had not yet attained classic endothermy to allow sustained rapid growth to adult body size (Chinsamy et al. 1994, 1995). More recently, Starck and I (2002) have proposed that perhaps basal birds inherited the flexible growth strategy from their dinosaurian ancestors and that perhaps in their later evolution they lost the ability for such flexibility in growth in response to selection pressure to grow rapidly to adult size within a single year. This would explain why large island birds, such as the 2 m tall Eocene *Diatryma* and the 3 m tall moa *Diornis* without strong pressure to grow rapidly, retain the pleisiomorphic flexible growth strategies like their nonavian dinosaur ancestors (see Fig. 6.6).

Growth Plates of Dinosaurs

Until fairly recently it was only cortical bone that was involved in the debates surrounding dinosaur thermal biology. However, now even the growth plate of dinosaur long bones (i.e., the chondro-osseous junction) has entered the debate. Claudia Barreto and her colleagues (1993) studied the cartilaginous growth plate of an adult monitor lizard, an 8–10-month-old dog, a 2-week-old chick, and a hatchling *Maiasaura*. They deduced that unlike mammals and lizards, where the chondro-osseous junction is straight, *Maiasaura* juveniles were similar to birds in having an undulating junction. As a result of this finding, they reasoned that dinosaurs, like birds, experienced a determinate growth strategy, which involved rapid growth facilitated by an endothermic physiology. A major problem with this study is that the experimental design was flawed in that it failed to include a disciplined comparison of the growth plates of juvenile / hatchling lizards such as *Varanus, Agama,* and *Chamaeleo.* In these forms, young individuals have well-developed cartilagenous canals and irregular growth plates similar to those seen in newborn mammals as well. These have been exceptionally well researched by Haines in the 1940s. Thus the reported similarities may result from inappropriate comparison of individuals at different stages of ontogeny, rather than indicating phylogenetic or physiological similarity. Hence, the basis for the fast growth rates and endothermic hypothesis is invalidated.

More recently, Horner and colleagues (2001) assessed embryonic archosaur bone microstructure and much emphasis was placed on the fact that apparent cartilage canals were present in the epiphyseal region of the embryonic and perinatal dinosaurs. I am skeptical that these are indeed cartilage canals, and feel that too much emphasis is placed on structures, which are indistinctive. Even so, cartilage canals are not only a characteristic of rapidly growing birds and mammals; they are also known from epiphyseal regions of *Varanus* bones (Haines 1942) and may indeed simply reflect rapid rates of longitudinal growth.

Dinosaur Lungs and Ventilation

As with most soft tissue, lungs of dinosaurs are not preserved. However, considering the extant phylogenetic bracket because extant crocodilians and birds have septate lungs (which is equivalent to a single oversized alveolus of a mammalian lung), it is reasonable to assume that dinosaurs also had septate lungs. However, the septate lungs of birds are highly derived structures that overcome the gas exchange limitations of the more typical reptilian septate lung. In birds, septal ingrowths, termed the *ediculae* or *faveoli* (where gaseous exchange occurs), are highly modified to form parabronchi, which boost the anatomical diffusion factor (ADF) to levels that generally exceed that of mammals. (ADF is an index to compare maximal gas exchange by considering the ratio of pulmonary respiratory surface area to mean blood gas barrier.) Reptiles, however, have fairly unspecialized septate lungs, with comparatively low levels of ADF. Among crocodilians, though, a modified heterogeneous, multicameral septate lung is present, with basic modifications toward the avian-type lung (i.e., heterogenous chambers, tubular dorsal chambers, and communication pores between adjacent chambers).

With regard to ventilation of lungs, extant crocodilians and birds also differ in the way they ventilate their lungs. Reptiles typically use a diaphragm and costal ventilation. However, crocodilians have a nonmuscular diaphragm that consists of a sheet of connective tissue that adheres to the cranial surface of the liver. Contraction of longitudinal muscles that extend between the liver, pubic bones, and caudal gastralia cause inspiratory movement that results in a pistonlike displacement of the liver-diaphragm complex that inflates the lungs. Thus they have what is referred to as a *hepatic-piston ventilatory mechanism* to aid ventilation of the lungs. Birds do not have a diaphragm and use costal ventilation and a system of air sacs (that act as bellows) to ventilate the lungs.

John Ruben and his colleagues (1996) see a separation between the relatively clear pleuropericardial and dark abdominal cavity in *Sinosauropteryx* (Fig. 6.3) and *Scipionyx* (Fig. 1.9) as suggesting the presence of a diaphragm. They suggested that these dinosaurs had a ventilatory mechanism unlike birds, which do not have a diaphragm, and more like reptiles. In *Scipionyx,* these researchers were able to locate the presence of longitudinal muscles on the pubis, which are in the same relative position in crocodiles as a structure they identify as the liver. Given that the positions of these structures are virtually identical to those of crocodiles, these researchers propose that they probably functioned in the same manner (i.e., for the hepatic-piston ventilation of the lungs). Ruben and his collaborators suggest that dinosaurs must have had septate lungs like those of crocodiles, which were ventilated by a similar hepatic-piston mechanism.

Considering that the basic attributes of a "protoparabronchial" lung are present in extant crocodilians and were possibly present in basal archosaurs, although there is no hard evidence, it is conceivable that in the evolution of dinosaurs, their lungs were evolving toward the condition in their descendants, extant birds. Perhaps new finds will shed more light onto respiratory structures and ventilation mechanisms of nonavian dinosaurs.

Nasal Turbinates

Among extant amniotes olfactory and respiratory turbinates occur in the nasal cavity. Olfactory turbinates, as the name suggests, are for smell, and are found among all amniotes. Respiratory turbinates are present in mammals and birds and appear to have evolved in tandem with high ventilatory rates, which result in large volumes of warm, humid air expelled from the lungs. Respiratory turbinates are generally located in the anterior part of the nasal cavity and in the path of expiratory air from which they absorb moisture and heat, which would otherwise lead to heat loss and desiccation. Thus, John Ruben and his colleagues (1997b) proposed that the absence or presence of nasal turbinates in dinosaurs is a good indication of minimal routine ventilatory rates. Three dinosaurs, *Nanotyrannus, Ornithomimus,* and *Hypacrosaurus,* skulls were CT scanned, but no respiratory turbinates were located. A few years ago, Bob Bakker had reported nasal turbinates in *Nanotyrannus,* but Ruben and colleagues (1997b) suggest that these are olfactory turbinates because they are located in the posterior part of the nasal cavity adjacent to the olfactory region of the brain.

Ruben and colleagues (1997b) also found that the nasal cavities of the dinosaurs were all fairly long and narrow and when plotted on a graph depicting cross-sectional area of the nares against body size, all three dinosaurs fell close to the ectotherm curve (Fig. 7.5). Recent work by Witmer (2001) on the position of the fleshy nostrils at the front of the skull (Fig. 2.3) seems to suggest that the nasal passages may have been longer than previously supposed and would therefore have aided the absorption of moisture and heat from air that traveled in and out of the nares. In the absence of nasal turbinates, it is questionable whether this is indicative of the same level of water recovery as in extant mammals and birds.

Thus on the basis of the absence of respiratory turbinates, it appears that at least some dinosaurs had a comparatively low rate of lung ventilation, which was probably more ectotherm-like. However, the dinosaurian capacity for maximum ventilatory rates is uncertain from the evidence at hand.

Curiously, unlike nonornithurine birds and nonavian dinosaurs, Cretaceous ornithurine birds have enlarged nasal chambers that are almost indistinguishable from those of extant birds. Jaap Hillenius and John Ruben (2004) suggest that it is likely that these birds had respiratory turbinates and probably had metabolic rates close to those of modern birds. Furthermore, it appears that Cretaceous ornithurine birds show specialized ventilatory mechanisms that are associated with extant birds: they have hinged

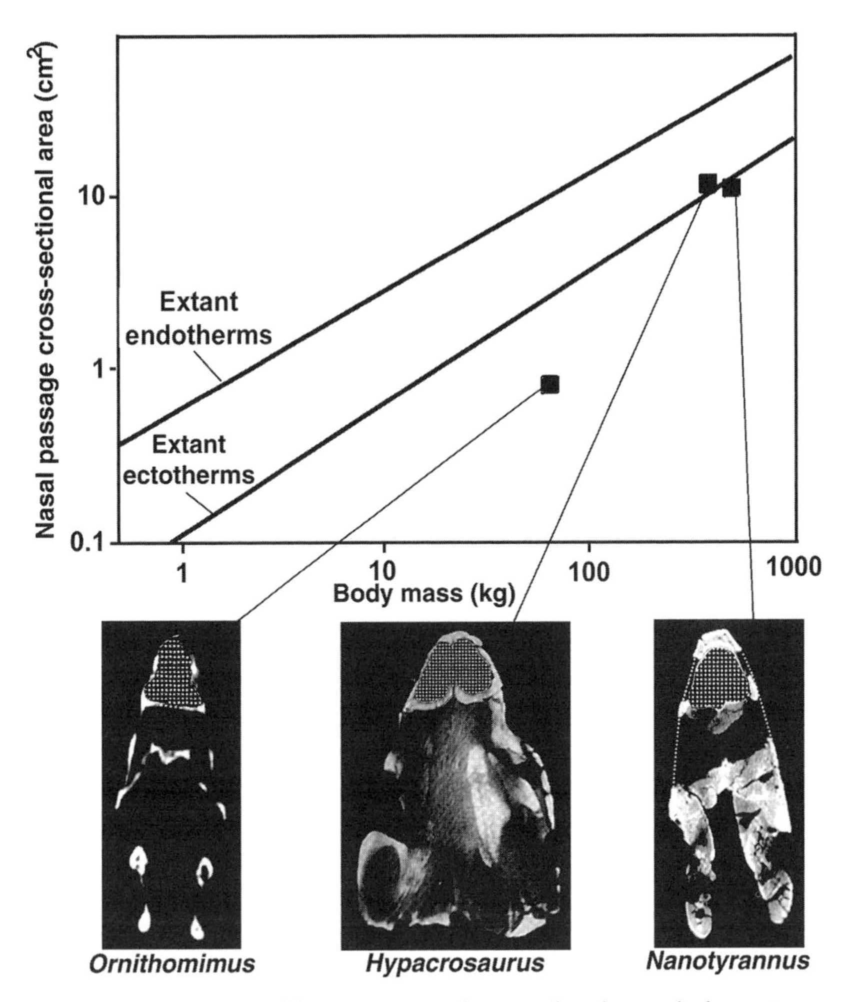

Fig. 7.5. Cross-sectional area of the respiratory nasal passage plotted against body mass in extant endotherms (birds and mammals) and ectotherms (lizards and crocodilians) and the Late Cretaceous dinosaurs, the hadrosaurid *Hypacrosaurus,* and theropods *Nanotyrannus* and *Ornithomimus.* Stippled areas in the inset CT cross sections through the rostrum indicate the nasal passages. Modified from Ruben et al. (1996). Courtesy of Willem Hillenius.

thoracic and sternal ribs, as well as transversely expanded sternocostal joints as in the ribcage of modern birds (Hillenius and Ruben 2000). Such structures do not occur in nonornithurine birds, nor are they present in nonavian dinosaurs. These findings suggest that endothermy, and its associated high metabolic rates, only evolved late in the evolution of dinosaurs among the Ornithurae.

Oxygen Isotopes

This method of deducing the physiology of dinosaurs relies on the assumption that the water the animal ingests is later chemically incorporated in the bone mineral of the animal and is dependent on the body temperature of the animal (e.g., Barrick and Showers 1995). Thus, by examining the ratio of O^{18}/O^{16} within a skeleton of an individual one would be able to deduce whether the animal was an ectotherm or an endotherm. It was reasoned that endotherms because of homeothermy would show relatively the same value irrespective of which bone in the skeleton is sampled. On the other hand, ectotherms would show more variability. Needless to say, this reasoning is seriously flawed because not only endotherms can attain homeothermy. It also appears that the so-called "homeothermic" signals obtained from dinosaurs are in fact diagenetic (i.e., the oxygen isotopes have been altered and are not a true reflection of the original condition that was present in the bones when the animal was alive). Researchers in Israel have found that the isotopic signal reflects latitudinal differences and also found that reptiles and mammals from the same deposits have homogenous signals (Kolodny and Luz 1993). It therefore appears that oxygen isotopes are not a reliable method to use to assess the thermal biology of extinct animals. This may change in the future if the methodology is refined to exclude diagenetic signals.

Conclusions

Each of the lines of evidence used to ascertain dinosaur physiology is not as straightforward as we would like them to be. In the case of bone microstructure, new work is suggesting that the bone growth is dependent on a number of factors and that bone tissue types cannot be easily linked to metabolic rates and physiology. Fibrolamellar bone is now known to occur quite widely among a variety of vertebrates, and its presence cannot be taken to signal high metabolic rates and endothermy. The same applies to Haversian bone. Indeed, it is appreciated that fibrolamellar bone forms at a faster rate than say lamellar bone in a particular taxon, but given the variability that exists with regard to bone depositional rates, and the fact that similar bone tissues do not mean similar bone depositional rates, caution is advised in the use of such actualistic hypotheses. Among the Mesozoic birds, it is evident that only ornithurine birds are able to grow at sustained, rapid uninterrupted rates, while all nonornithurine birds studied appear to have interrupted rates of growth and grew more slowly than their modern relatives (Chinsamy et al. 1994; Chinsamy, Martin, and Dodson 1998; Starck and Chinsamy 2001). It is interesting that only the ornithurine birds have skeletal evidence of a ventilatory mechanism as in extant birds, as well as wide nasal sinuses that prob-

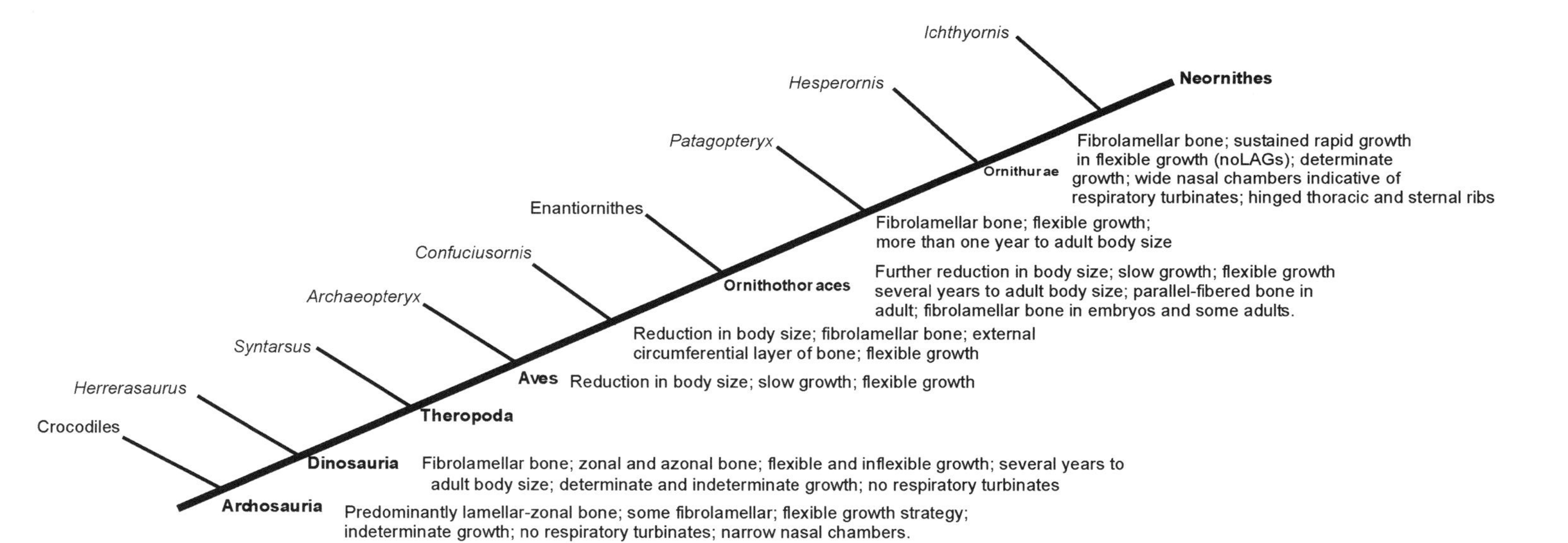

Fig. 7.6. Various characteristics of bone microstructure, lung structure, and ventilation plotted out onto a phylogeny of dinosaurs.

ably housed nasal turbinates, which are interpreted as suggestive that endothermy and its associated high levels of metabolism evolved fairly late in the evolution of dinosaurs (Hillenius and Ruben 2004). Perhaps, in tandem with this, the ability to grow rapidly at uninterrupted rates also evolved.

It turns out that until now there is no unambiguous evidence that supports a tachymetabolic endothermic physiology for any nonavian dinosaur (Chinsamy and Hillenius 2004). This does not necessarily mean that dinosaurs were sluggish, slow-moving, overgrown lizards. Given the large size of many dinosaurs, and considering the various models proposed regarding thermoregulation in large animals, it appears that endothermy need not be invoked for these animals to be homeothermic because regardless of metabolic capacity, large-bodied animals are able to maintain stable internal body temperatures. According to these models, small dinosaurs (i.e., less than 75 kg body weight) would have been affected by temperature fluctuations, but moderate to large-sized dinosaurs (greater than 500 kg) would have been able to use their bulk to escape temperature variation. Thus, according to physiologists Jim Spotila and Frank Paladino, no special adaptations except large body size would be needed to achieve homeothermy (a physiological specialization termed *gigantothermy*). Considering that the Mesozoic climate was relatively warm and equable, even the smaller dinosaurs could have used behavioral thermoregulation to maintain homeothermy. Thus, although there is no hard evidence so far to support proposals that dinosaurs were endothermic, dinosaurs could still have been active and dynamic animals in their Mesozoic world.

In an attempt to understand the direction of evolutionary change, the various characteristics used to infer physiology are mapped out on a well-established phylogeny of the Dinosauria (Fig. 7.6). From this it is evident that in the evolution of the Dinosauria distinct trends in growth rates can be observed. The basal archosaurs have a slow, flexible growth, while the Dinosauria have a faster growth rate that was both flexible and inflexible. Basal nonornithurine birds also show plasticity in their growth, although they generally have a fast growth rate like their nonavian ancestors (e.g., Chinsamy et al. 1994). The Ornithurae are distinctive in having the fastest growth rates, as reflected by their uninterrupted cortical bone (i.e., lack of plasticity; Chinsamy, Martin, and Dodson 1998; Chinsamy 2002). Thus, in the evolution of the Archosauria, there appears to be an increasing trend toward faster growth. However, only in the Ornithurae where the fastest growth rates are evident, signals of higher ventilation rates also occur (such as wider narial chambers and hinged thorasic and sternal ribs). It is likely that rapid rates of growth (as deduced from bone microstructure) and high ventilation rates (as suggested from the wide nasal sinuses) probably occurred in tandem with the evolution of high metabolic rates and endothermy in the Ornithurae.

References

Amprino, R. 1947. La structure du tissu osseux envisagée comme expression de différences dans la vitesse de l'accroissement. *Archives de Biologie,* 58, 315–330.

Amprino, R., and Godina, G. 1947. La struttura della ossa nei vertebrati: Ricerche comparative negli amfibi e negli amnioti. *Commentationes Pontificiae Academiae Scientiarum,* 11, 329–467.

———. 1967. Bone histophysiology. *Guys Hospital Report,* 116, 51–69.

Anderson, J. F., Hall-Martin, A., and Russel, D. A. 1985. Long bone circumference and weight in mammals, birds and dinosaurs. *Journal of Zoology, London,* 207, 53–61.

Austin, J. J., Smith, A. B., and Thomas, R. H. 1997. Palaeontology in a molecular world: The search for authentic ancient DNA. *Trends in Ecology and Evolution,* 12(8), 303–306.

Bakker, R. T. 1971. Dinosaur physiology and the origin of mammals. *Evolution,* 25(4), 636–658.

———. 1972. Anatomical and ecological evidence of endothermy in dinosaurs. *Nature,* 238, 81–85.

———. 1975. Dinosaur renaissance. *Scientific American,* 232, 58–78.

———. 1986. *The Dinosaur Heresies: New Theories Unlocking the Mystery of the Dinosaurs and Their Extinction.* William Morrow, New York.

Barreto, C., Albrecht, R. M., Bjorling, D. E., Horner, J. R., and Wilsman, N. J. 1993. Evidence of growth plate and the growth of long bones in juvenile dinosaurs. *Science,* 264, 2020–2023.

Barrick, R. E., and Showers, W. J. 1995. Oxygen isotope variability in juvenile dinosaurs (*Hypacrosaurus*): Evidence for thermoregulation. *Paleobiology,* 21(4), 552–560.

Behrensmeyer, A. K., and Hill, A. P. 1980. *Fossils in the Making: Vertebrate Taphonomy and Paleoecology.* University of Chicago Press, Chicago.

Benton, M. J. 1989. Evolution of large size. In *Palaeobiology: A Synthesis,* ed. D.E.G. Briggs and P. R. Crowther, 147–152. Blackwell, Oxford.

Bertram, J. E. A., and Swartz, S. M. 1991. The "Law of bone transformation": A case of crying Wolff? *Biological Reviews,* 66, 245–273.

Botha, J. 2001. The Paleobiology of the Non-mammalian Cynodonts Deduced from Bone Microstructure and Stable Isotopes. PhD thesis, University of Cape Town.

Botha, J., and Chinsamy, A. 2000. Growth patterns deduced from the bone histology of the Cynodonts *Diademodon* and *Cynognathus. Journal of Vertebrate Paleontology,* 20(4), 705–711.

———. 2004. Growth and life habits of the Triassic cynodont *Trirachodon,* inferred from bone histology. *Acta Palaeontologia Polonica,* 49, 619–627.

Briggs D. E. G., Wilby, P. R., Moreno, B. P. P., Sanz, J. L., and Martínez, M. F. 1997. The mineralization of dinosaur soft tissue in the Lower Cretaceous of Las Hoyas, Spain. *Journal of the Geological Society, London,* 154, 587–588.

Brinkman, D. L., and Conway, F. M. 1985. Textural and mineralogical analysis of a *Stegosaurus* (Reptilia: Ornithischia) plate. *Compass,* 63(1), 1–5.

Brochu, C. A. 1996. Closure of neurocentral sutures during crocodilian ontogeny: Implications for maturity assessment in fossil archosaurs. *Journal of Vertebrate Paleontology,* 16, 49–62.

Broili, F. 1922. Ueber den feineren Bau der "verknöcherten Sehnen" (=verknöcherten Muskeln) von *Trachodon. Anatomischer Anzeiger,* 55, 465–475.

Buffrénil, V. de, Farlow, J. O., and Ricqlès, A. de. 1986. Growth and function of *Stegosaurus* plates: Evidence from bone histology. *Paleobiology,* 12(4), 459–473.

Buffrénil, V. de, and Mazin, J. M. 1989. Bone histology of *Claudiosaurus germaini* (Reptilia, Claudiosauridae) and the problem of pachyostosis in aquatic tetrapods. *Historical Biology,* 2, 311–322.

Buffrénil, V. de, and Shoevaert, D. 1988. On how the delphinid humerus becomes cancellous: Ontogeny of a histological specialization. *Journal of Morphology,* 198, 149–164.

———. 1989. Données quantitatives et observations histologiques sur la "pachyostose" du squellette du Dugong, *Dugong dugong* (Müller) (Sirenia, Dugongidae). *Canadian Journal of Zoology,* 67, 2107–2119.

Bühler, P. 1986. Das Vogelskelett-hochentwickelter Knochen-leichtbau. *Arcus,* 5, 221–228.

Caetano, M. H. 1990. Use and results of skeletochronology in some urodeles (*Triturus marmoratus,* Latreille 1800 and *Triturus boscai,* Lataste 1879. *Annales des Sciences Naturelles, Zoologie,* 11, 197–199.

Campbell, B., and Lack, E. 1985. *A Dictionary of Birds.* T & AD Poyser, Calton.

Campbell, J. G. 1966. A dinosaur bone lesion resembling avian osteopetrosis and some remarks on the mode of development of the lesions. *Journal of the Royal Microscopical Society,* 85, 163–174.

Carter, D. R., Mikic, B., and Padian, K. 1998. Epigenetic mechanical factors in the evolution of long bone epiphyses. *Zoological Journal of the Linnean Society,* 123, 163–178.

Carter, D. R., and Spengler, D. M. 1978. Mechanical properties and composition of cortical bone. *Clinical Orthopaedics and Related Research,* 135, 192–217.

Carter, T. J. 1978. Speculations on the growth rate and reproduction of some dinosaurs. *Paleobiology,* 4(3), 320–328.

Casinos, A., Quintana, C., and Viladiu, C. 1993. Allometry and adaptation in the long bones of a digging group of rodents (Ctenomyinae). *Zoological Journal of the Linnean Society,* 107, 107–115.

Castanet, J., and Baez, M. 1991. Adaptation and evolution in Gallotia lizard from the Canary Islands: Age, growth, maturity and longevity. *Amphibia-Reptilia,* 12, 81–102.

Castanet, J., Grandin, A., Abourachid, A., and Ricqlès, A. de. 1996. Expression de la dynamique de croissance dans la structure de l'os périostique chez *Anas platyrhynchos. Life Sciences,* 319, 301–308.

Castanet, J., Rogers, K. R., Cubo, J., and Boisard, J. 2000. Periosteal bone growth rates in extant ratites (ostriche and emu). Implications for assessing growth in dinosaurs. *Life Sciences,* 323, 543–550.

Castanet, J., and Smirina, E. 1990. Introduction to the skeletochronological method in amphibians and reptiles. *Annales des Sciences Naturelles, Zoologie,* 11, 191–196.

Castanet, J., Vieillot, H. F., Meunier, F. J., and Ricqlès, A. de. 1993. Bone and individual aging. In *Bone.* Vol. 7, *Bone Growth,* ed. B. K. Hall, 245–283. CRC Press, Boca Raton, FL.

Chen, P.-J., Dong, Z.-M., and Zhen, S.-N. 1998. An exceptionally well-preserved theropod dinosaur from the Yixian Formation of China. *Nature,* 391, 147–152.

Chiappe, L. M. 1995. The first 85 million years of avian evolution. *Nature,* 378, 349–355.

Chiappe, L. M., Coria, R. A., Dingus, L., Jackson, F., Chinsamy, A., and Fox, M. 1998. Sauropod dinosaur embryos from the Late Cretaceous of Patagonia. *Nature,* 396, 258–261.

Chiappe, L. M., and Dyke, G. J. 2002. The Mesozoic radiation of birds. *Annual Review of Ecology and Systematics,* 33, 91–124.

Chiappe, L. M., Shu'an, J., Qiang, J., and Norell, M. A. 1999. Anatomy and systematics of the Confusiusornithidae (Theropoda: Aves) from the Late Mesozoic of Northeastern China. *Bulletin of the American Museum of Natural History,* 242, 1–89.

Chiappe, L. M., and Witmer, L. M. 2001. *Mesozoic birds: Above the Heads of Dinosaurs.* University of California Press, Berkeley.

Chinsamy, A. 1990. Physiological implications of the bone histology of *Syntarsus rhodesiensis* (Saurischia: Theropoda). *Palaeontologia africana,* 27, 77–82.

———. 1991. The Osteohistology of Femoral Growth within a Clade: A Comparison of the Crocodile, Crocodylus niloticus, the Dinosaurs, Massospondylus and Syntarsus, and the Birds, Struthio and Sagittarius. PhD thesis. University of the Witwatersrand, Johannesburg.

———. 1993a. Bone histology and growth trajectory of the prosauropod dinosaur *Massospondylus carinatus* (Owen). *Modern Geology,* 18, 319–329.

———. 1993b. Image analysis and the physiological implications of the vascularisation of femora in archosaurs. *Modern Geology,* 19, 101–108.

———. 1994. Dinosaur bone histology: Implications and inferences. In *Dinofest,* ed. G. D. Rosenberg and D. Wolberg, 213–227. Special Publication No. 7. Paleontological Society, University of Tennessee, Knoxville.

———. 1995a. Histological perspectives on growth in the birds *Struthio camelus* and *Sagittarius serpentarius. Courier Forschungsinstitut Senckenberg,* 181, 317–323.

———. 1995b. Ontogenetic changes in the bone histology of the Late Jurassic ornithopod *Dryosaurus lettowvorbecki. Journal of Vertebrate Palaeontology,* 15(3), 96–104.

———. 1997. Assessing the biology of fossil vertebrates through bone histology. *Palaeontologia africana,* 33, 29–35.

———. 2002. Bone microstructure of early birds. In *Mesozoic Birds: Above the Heads of Dinosaurs,* ed. L. M. Chiappe and L. M. Witmer, 421–431. University of California Press, Berkeley.

Chinsamy, A., and Barrett, P. M. 1997. Sex and old bones. *Journal of Vertebrate Paleontology,* 17, 450.

Chinsamy, A., Chiappe, L., and Dodson, P. 1994. Growth rings in Mesozoic avian bones: Physiological implications for basal birds. *Nature,* 368, 196–197.

———. 1995. Mesozoic avian bone microstructure: Physiological implications. *Paleobiology,* 21, 561–574.

Chinsamy, A., and Dodson, P. 1995. Inside a dinosaur bone. *American Scientist,* 83, 174–180.

Chinsamy, A., and Elzanowski, A. 2001. Evolution of growth pattern in birds. *Nature,* 412, 402–403.

Chinsamy, A., February, E. C., Harley, E., Rich, T. H., and Vickers, P. R. 1996. Wood within a polar dinosaur bone. *South African Journal of Science,* 92, 48.

Chinsamy, A., Hanrahan, S. A., Neto, R. M., and Seely, Y. M. 1995. Skeletochronological assessment of age in *Angolosaurus skoogi,* a Cordylid lizard living in an aseasonal environment. *Journal of Herpetology,* 29(3), 457–460.

Chinsamy, A., and Hillenius, W. J. 2004. Physiology of non-avian dinosaurs. In *The Dinosauria.* 2nd ed., ed. D. B. Weishampel, P. Dodson, and H. Osmólska, 643–659. University of California Press, Berkeley.

Chinsamy, A., Martin, L. D., and Dodson, P. 1998. Bone microstructure of diving *Hesperornis* and the volant *Ichthyornis* from the Niobrara Chalk of western Kansas. *Cretaceous Research,* 19, 225–235.

Chinsamy, A., and Raath, M. A. 1992. Preparation of bone for histological study. *Palaeontologia africana,* 29, 39–44.

Chinsamy, A., Rich, T., and Vickers-Rich, P. 1998. Polar dinosaur bone histology. *Journal of Vertebrate Paleontology,* 18, 385–390.

Chinsamy, A., and Rubidge, B. 1993. Dicynodont (Therapsida) bone histology: Phylogenetic and physiological implications. *Palaeontologia africana,* 30, 97–102.

Clarke, J. B., and Barker, M. J. 1993. Diagenesis in *Iguanodon* bones from the Wealden Group, Isle of Wight, Southern England. *Kaupia,* 2, 57–65.

Constantine, A., Chinsamy, A., Vickers-Rich, P., and Rich, T. H. 1998. Periglacial environments and polar dinosaurs. *South African Journal of Science,* 94, 137–141.

Cook, S. F., Brooks, S. T., and Ezra-Cohn, H. E. 1962. Histological studies on fossil bone. *Journal of Paleontology,* 36(3), 483–494.

Coombs, W. P. 1990. Behavior patterns of dinosaurs. In *The Dinosauria,* ed. D. B. Weishampel, P. Dodson, and H. Osmólska, 32–42. University of California Press, Berkeley.

Cracraft, J. 2001. Avian evolution: Gondwana biogeography and the Cretaceous-Tertiary mass extinction event. *Proceedings of the Royal Society London,* 268, 459–469.

Crompton, A. W. 1989. Bone remodeling in medium-sized mammals and birds in response to slight changes in normal strain levels. *Journal of Vertebrate Paleontology,* 10(4), 18A.

Currey, J. D. 1960. Differences in the blood supply of bone of different histological types. *Quarterly Journal of Microscopical Science,* 101, 351–370.

———. 1962. The histology of the bone of a Prosauropod dinosaur. *Palaeontology,* 5(2), 238–246.

———. 1984. Effects of differences in mineralization on the mechanical properties of bone. *Philosophical Transactions of the Royal Society of London B,* 304, 509–518.

———. 1991. Biomechanics of mineralized skeletons. In *Patterns, Processes and Evolutionary Trends,* ed. J. G. Carter, 11–25. Van Nostrand Reinhold, New York.

Currey, J. D., and Alexander, M. 1985. The thickness of the walls of tubular bones. *Journal of Zoology (London),* 206, 453–468.

Currey, K. A. 1999. Ontogenetic histology of *Apatosaurus* (Dinosauria: Sauropoda): New insights on growth rates and longevity. *Journal of Vertebrate Paleontology,* 19, 654–665.

De Klerk, W. J., Forster, C. A., Sampson, S. D., Chinsamy, A., and Ross, C. F. 2000. A new coelurosaurian dinosaur from the Early Cretaceous of South Africa. *Journal of Vertebrate Paleontology,* 20, 324–332.

Dollo, L. 1886. Note sur les ligaments ossifiés des dinosauriens de Bernissart. *Archs Biol.,* 7, 249–264.

Dunham, A. E., Overall, K. L., Porter, W. P., and Forster, C. A. 1989. Implications of ecological energies and biophysical and developmental constraints for life-history variation in dinosaurs. *Geological Society of America Special Paper,* 238, 1–19.

Eggling, H. von. 1938. Allegemeines über den aufbau knöcherner skeleteile. In *Handbuch für vergleichenden Anatomie der Wirbeltiere,* ed. L. V. Bolk, E. Göppert, E. Kallius, and W. Lubosch, 275–304. Urban et Schwarzenberg, Berlin.

Elzanowski, A. 1981. Embryonic bird skeletons from the Late Cretaceous of Mongolia. *Palaeontologia Polonica,* 42, 147–179.

Enlow, D. H. 1962a. Functions of the Haversian system. *American Journal of Anatomy,* 110, 268–306.

———. 1962b. A study of the post-natal growth and remodeling of bone. *American Journal of Anatomy,* 110, 79–102.

———. 1963. *Principles of Bone Remodeling.* Charles C Thomas, Springfield, IL.

———. 1969. The bone of reptiles. In *Biology of the Reptilia: Morphology A,* ed. C. Gans and A. d'A. Bellairs, 45–77. Academic Press, London.

Enlow, D. H., and Brown, S. O. 1956. A comparative histological study of fossil and recent bone tissue. Pt. 1. *Texas Journal of Science,* 8, 405–443.

———. 1957. A comparative histological study of fossil and recent bone tissue. Pt. 2. *Texas Journal of Science,* 9, 186–214.

———. 1958. A comparative histological study of fossil and recent bone tissue. Pts. 2 & 3. *Texas Journal of Science,* 10, 187–230.

Erickson, G. M. 1999. Breathing life into *Tyrannosaurus rex. Scientific American,* 34–39.

Erickson, G. M., Makovicky, P. J., Currie, P. J., Norell, M. A., Yerby, S. A., and Brochu, C. A. 2004. Gigantism and comparative life-history parameters of tyrannosaurid dinosaurs. *Nature,* 430, 772–775.

Erickson, G. M., Rogers, K. C., and Yerby, S. A. 2001. Dinosaurian growth patterns and rapid avian growth rates. *Nature,* 412, 429–433.

Erickson, G. M., and Tumanova, T. A. 2000. Growth curve of *Psittacosaurus mongoliensis* Osborn (Ceratopsia: Psittacosauridae) inferred from long bone histology. *Zoological Journal of the Linnaean Society,* 130, 551–566.

Erickson, G. M., Van Kirk, S. D., Su, J., Levenston, M. E., Caler, W. E., and Carter, D. R. 1996. Bite-force estimation for *Tyrannosaurus rex* from tooth-marked bones. *Nature,* 382, 706–708.

Farlow, J. O. 1990. Dinosaur energetics and thermal biology. In *The Dinosauria,* ed. D. B. Weishampel, P. Dodson, and H. Osmólska, 43–55. University of California Press, Berkeley.

———. 1993. On the rareness of big, fierce animals: Speculations about the body sizes, population densities, and geographic ranges of predatory mammals and large carnivorous dinosaurs. *American Journal of Science,* 293–A, 167–199.

Farlow, J. O., Dodson, P., and Chinsamy, A. 1995. Dinosaur biology. *Annual Review of Ecology and Systematics,* 26, 445–471.

Feduccia, A. 1999. *The Origin and Evolution of Birds.* Yale University Press, New Haven, CT.

Finch, C. A., Brown, E. B., Brown, J. C., Bumsted, H. E., Bundy, M., Hem, J. D., Neilands, et al. 1979. *Iron.* University Park Press, Baltimore.

Fisher, P. E., Russell, D. A., Stoskopf, M. K., Barrick, R. E., Hammer, M., and Kuzmitz, A. A. 2000. Cardiovascular evidence for an intermediate or higher metabolic rate in an Ornithischian Dinosaur. *Science,* 288, 503–505.

Foote, J. S. 1916. A contribution to the comparative histology of the femur. *Smithsonian Contributions to Knowledge,* 35, 1–242.

Forster, C. A., Chiappe, L. M., Krause, D. W., and Sampson, S. D. 1996. The first Cretaceous bird from Madagascar. *Nature,* 382, 532–534.

Francillon-Vieillot, H., Buffrénil, V. de, Castanet, J., Géraudie J., Meunier, F. J., Sire, J. Y., Zylberberg, L., and Ricqlès, A. de. 1990. Microstructure and mineralization of vertebrate skeletal tissues. In *Skeletal Biomineralisation: Patterns, Processes and Evolutionary Trends,* ed. J. G. Carter, 471–530. Van Nostrand Reinhold, New York.

Fukuda, Y., and Obata, I. 1993. Discovery of osteocytes in adult and juvenile bones of duck-billed dinosaurs. *Natural History Research,* 2, 99–102.

Garland, A. N. 1987. A histological study of archaeological bone decomposition. In *Death, Decay, and Reconstruction: Approaches to Archaeology and Forensic Science,* ed. A. Boddington, A. N. Garland, and R. C. Janaway, 109–126. University Press, Manchester.

Geist, N. R., and Jones, T. D. 1996. Juvenile skeletal structure and the reproductive habits of dinosaurs. *Science,* 272, 712–714.

Geist, N. R., Jones, T. D., and Ruben, J. A. 1997. Implications of soft tissue preservation in the compsognathid dinosaur, *Sinosauropteryx. Journal of Vertebrate Paleontology,* 17, 48A.

Golenberg, E. M., Giannassi, D. E., Clegg, M. T., Smiley, C. J., Durbin, M., Henderson, D., and Zurawski, G. 1990. Chloroplast DNA from a Miocene *Magnolia* species. *Nature,* 344, 656–658.

Gross, W. 1934. Die Typen des mikroskopischen Knochenbaues bei fossilen Stegocephalen und Reptilien. *Zeitschrift für Anatomie,* 103, 731–764.

Gurley, L. R.,Valdez, J. G., Spall, W. D., Smith, B. F., and Gillette, D. D. 1991. Proteins in the fossil bone of the dinosaur. *Seismosaurus. Journal of Protein Chemistry,* 10(1).

Haines, R. W. 1942. The evolution of epiphysis and of endochondral bone. *Biological Reviews,* 174, 267–292.

———. 1969. Epiphyses and sesamoids. In *Biology of the Reptilia,* ed. C. Gans and A. d'A. Bellairs, 81–115. Academic Press, New York

Ham, A. W. 1953. *Histology.* 2nd ed. Lippincott, Philadelphia.

Hasse, C. 1878. Die fossilen Wirbel. Morphologische Studien. Die Histologie fossiler reptilwirbel. *Morphologishe Jahrbuch,* 4, 480–502.

Heilmann, G. 1926. *Origin of Birds.* Witherby, London.

Heinrich, R. E., Ruff, C. B., and Weishampel, D. B. 1993. Femoral ontogeny and locomotor biomechanics of *Dryosaurus lettowvorbecki* (Dinosauria, Iguanodontia). *Zoological Journal of the Linnaean Society,* 108, 176–196.

Hillenius, W. J., and Ruben, J. 2004. The evolution of endothermy in terrestrial vertebrates: Who? When? Why? *Physiological and Biochemical Zoology,* 77, 1019–1042.

Hodge, A. J., and Petruska, J. A. 1963. Recent studies with the electron microscope on ordered aggregates of the tropocollagen molecule. In *Aspects of Protein Structure,* ed. G. N. Ramachandran, 289–300. Academic Press, New York.

Horner, J. R., and Currie, P. H. 1994. Embryonic and neonatal morphology and ontogeny of a new species of *Hypacrosaurus* (Ornithischia, Lambeosauridae) from Montana and Alberta. In *Dinosaur Eggs and Babies,* ed. K. Carpenter, K. F. Hirsch, and J. R. Horner, 312–332. Cambridge University Press, Cambridge.

Horner, J. R., and Makela, R. 1979. Nest of juveniles provides evidence of family structure among dinosaurs. *Nature,* 282, 296–298.

Horner, J. R., and Padian, K. 2004. Age and growth dynamics of *Tyrannosaurus rex. Proceedings of the Royal Society of London B,* 271, 1875–1880.

Horner, J. R., Padian, K., and Ricqlès, A. de. 2000. Long bone histology of the hadrosaurid dinosaur *Maiasaura peeblesorum:* Growth dynamics and physiology based on an ontogenetic series of skeletal elements. *Journal of Vertebrate Paleontology,* 20, 115–129.

———. 2001. Comparative osteohistology of some embryonic and perinatal archosaurs: Developmental and behavioral implications for dinosaurs. *Paleobiology,* 27, 39–58.

Horner, J. R., Ricqlès, A. de, and Padian, K. 1999a. Variation in dinosaur skeletochronology indicators: Implications for age assessment and physiology. *Paleobiology,* 25(3), 295–304.

Horner, J. R., and Weishampel, D. B. 1988. A comparative embryological study of two ornithischian dinosaurs. *Nature,* 332, 256–257.

Houde, P. 1987. Histological evidence for the systematic position of *Hesperornis* (Odontornithes: Hesperornithiformes). *Auk,* 104, 125–129.

Hubert, J. F., Parish, P. T., Chure, D. J., and Prostak, K. S. 1996. Chemistry, microstructure, petrology, and diagenetic model of Jurassic dinosaur bones, Dinosaur National Monument, Utah. *Journal of Sedimentary Research,* 66, 531–547.

Hutton, J. M. 1986. Age determination of living Nile crocodiles from the cortical stratification of bone. *Copeia,* 1986, 332–341.

Janensch, W. 1947. Pneutimatizität bei wirbeln von sauropoden und anderen saurischiern. *Palaeontographica,* 7, 1–25.

Ji, Q., Currie, P. J., Norell, M., and Ji, S.-A. 1998. Two feathered theropods from the Upper Jurassic/Lower Cretaceous strata of northeastern China. *Nature,* 393, 753–761.

Jones, T. D., Farlow, J. O., Ruben, J. A., Henderson, D. M., and Hillenius, W. J. 2000. Cursoriality in bipedal archosaurs. *Nature,* 406, 716–718.

Kellner, A. W. A. 1996. Fossilized theropod soft tissue. *Nature,* 379, 32.

Kielan-Jaworowska, Z., and Pensko, J. 1967. Unusually radioactive fossil bones from Mongolia. *Nature,* 214, 161–163.

Kitching, J. W. 1979. Preliminary report on a clutch of six dinosaurian eggs from the Upper Triassic Elliot Formation, northern Orange Free State. *Palaeontologia africana,* 22, 41–45.

Klevezal, G. A. 1996. *Recording Structures of Mammals: Determination of Age and Reconstruction of Life History.* Balkema, Rotterdam.

Klevezal, G. A., and Kleinenberg, S. E. 1969. *Age Determination of Mammals from Annual Layers in Teeth and Bones.* Translated from Russian by J. Salkind. Israel Program for Scientific Translations Press, Jerusalem.

Kolodny, Y., and Luz, B. 1993. Dinosaur thermal physiology from $\delta^{18}O$ in bone phosphate; is it possible? In *Second Oxford Workshop on Bone Diagenesis.* Abstracts. Oxford University Press, Oxford.

Krings, M., Stone, A., Schmitz, R. W., Krainitzki, H., Stoneking, M., and Pääbo, S. 1997. Neanderthal DNA sequences and the origin of modern humans. *Cell,* 90, 19–30.

Lacroix, P. 1971. The internal remodeling of bones. In *The Biochemistry and Physiology of Bone,* ed. G. H. Bourne, 119–144. Academic Press, New York.

Lapparent, A. F. de. 1947. Les Dinosauriens dur Crétacé supérieur du Midi de la France. *Mémoirs de la Société géologique de France (Nouvelle Série),* 56, 1–54.

Leahy, G. D. 1991. Lamellar-zonal bone in fossil mammals: Implications for dinosaur and therapsid paleophysiology. *Journal of Vertebrate Paleontology,* 11, 42A.

Lee Lyman, R. 1994. *Vertebrate Taphonomy.* Cambridge University Press, Cambridge.

Li, Y., An, C.-C., Zhu, Y.-X., Zhang, Y., Liu, Y.-F., Qu, L., You, L.-T., et al. 1995. DNA isolation and sequence analysis of dinosaur DNA from Cretaceous dinosaur egg in Xixia Henan, China. *Acta Scientiarum Naturalium Universitatis Pekinensis,* 31, 148–152.

Lieberman, D. E. 1996. How and why humans grow thin skulls: Experimental evidence for systematic cortical robusticity. *American Journal of Physical Anthropology,* 101, 217–236.

Lieberman, D. E., and Pearson, O. M. 2001. Trade-off between modeling and remodeling responses to loading in the mammalian limb. *Bulletin of the Museum of Comparative Zoology,* 156(1), 269–282.

Lincoln, R. J., Boxshall, G. A., and Clark, P. F. 1982. A dictionary of ecology, evolution and systematics. Cambridge, Cambridge University Press.

Lowenstam, H. A., and Weiner, S. 1989. *On Biomineralization.* Oxford University Press, New York.

Maderson, P. F. A., and Hillenius, W. J. 2004. Feather development and evolution: Only the whole story will suffice. *Journal of Morphology,* 260, 308.

Madsen, J. 1976. *Allosaurus fragilis:* A revised osteology. *Utah Geological and Mineral Survey Bulletin,* 109, 1–63.

Mantell, G. A. 1850. On *Pelorosaurus,* an undescribed gigantic terrestrial reptile, whose remains are associated with *Iguanodon* and other saurians in the strata of Tilgate Forest, Sussex. *Philosophical Transactions of the Royal Society,* 140, 379–390.

Margerie, E. de, Cubo, J., and Castanet, J. 2002. Bone typology and growth rate: Testing and quantifying "Amprino's rule" in the mallard (*Anas platyrhynchos*). *Compte Rendu Biologies,* 325, 221–230.

Margerie, E. de, Robin, J.-P., Verrier, D., Cubo, J., Groscolas, R., and Castanet, J. 2004. Assessing a relationship between bone microstructure and growth rate: A fluorescent labelling study in the king penguin chick (*Aptenodytes patagonicus*). *Journal of Experimental Biology,* 207, 869–879.

Martill, D. M. 1987. Taphonomic and diagenetic case study of a partially articulated Ichthyosaur. *Palaeontology,* 30, 543–555.

———. 1991. Bones as stones: The contribution of vertebrate remains to the lithologic record. In *The Process of Fossilization,* ed. S. K. Donovan, 271–292. Columbia University Press, New York.

Martill, D. M., Barker, M. J., and Dacke, C. G. 1996. Dinosaur nesting or preying? *Nature,* 379, 778.

Martill, D. M., and Unwin, D. M. 1997. Small spheres in fossil bones: Blood corpuscles or diagenetic products? *Palaeontology,* 40(3), 619–624.

Martin, B., and Burr, D. B. 1989. *Structure, Function, and Adaptation of Compact Bone.* Raven Press, New York.

Martin, L. D. 1983. The origin and early radiation of birds. In *Perspectives in Ornithology,* ed. A. H. Brush and G. A. Clark, 291–338. Cambridge University Press, New York.

McGowan, C. 1991. *Dinosaurs, Spitfires, and Sea Dragons.* Harvard University Press, Cambridge, MA.

Moodie, R. L. 1923. *Palaeopathology.* University of Illinois Press, Urbana.

———. 1928. The histological nature of ossified tendons found in dinosaurs. *American Museum Novitates,* 311, 1–15.

Muyzer, G., Sandberg, P., Knapen, M. H. J., Vermeer, C., Collins, M., and Westbroek, P. 1992. Preservation of the bone protein osteocalcin in dinosaurs. *Geology,* 20, 871–874.

Nopsca, F. von, and Heidsieck, E. 1933. On the bone histology of the ribs in immature and half-grown trachodont dinosaurs. *Proceedings of the Royal Zoological Society,* 1933, 221–223.

Norell, M., Ji, Q., Gao, K., Yuan, C., Zhao, Y., and Wang, L. 2002. "Modern" feathers on a non-avian dinosaur. *Nature,* 416, 36–37.

Norell, M. A., Clark, J. M., Chiappe, L. M., and Dashzeveg, D. 1995. A nesting dinosaur. *Nature,* 378, 774–776.

Norell, M. A., Makovicky, P., and Clark, J. M. 1997. A *Velociraptor* wishbone. *Nature,* 389, 447.

Ovchinnikov, I. V., Götherström, A., Romanoval, G. P., Kharitonov, V. M., Lidén, K., and Goodwin, W. 2000. Molecular analysis of Neanderthal DNA from the northern Caucasus. *Nature*, 404, 490–493.

Owen, R. 1859. Monograph on the fossil Reptilia of the Wealden and Purbeck Formations. *Palaeontographical Society Monographs*, Suppl. no. 2, 11, 20–44.

Padian, K. 1997. Physiology. In *Encyclopedia of Dinosaurs,* ed. P. J. Currie and K. Padian, 552–557. Academic Press, London.

Padian, K., and Chiappe, L. M. 1998. The origin and early evolution of birds. *Biological Reviews,* 73, 1–42.

Padian, K., Ricqlès, A. de, and Horner, J. R. 2001. Dinosaurian growth rates and bird origins. *Nature,* 412, 405–408.

Paladino, F. V., Spotila, J. R., and Dodson, P. 1997. A blueprint for giants: Modeling the physiology of large dinosaurs. In *The Complete Dinosaur,* ed. J. O. Farlow and M. K. Brett-Surman, 491–504. Indiana University Press, Bloomington.

Parker, R. B., and Toots, H. 1970. Minor elements in fossil bone. *Bulletin of the Geological Society of America,* 81, 925–932.

———. 1980. Trace elements in bones as paleobiological indicators. In *Fossils in the Making: Vertebrate Taphonomy and Paleoecology,* ed. A. K. Behrensmeyer and A. P. Hill, 197–207. University of Chicago Press, Chicago.

Patnaik, B. K., and Behera, M. N. 1981. Age determination in the tropical agamid lizard, *Calotes versicolor* (Daudin) based on bone histology. *Experimental Gerontology,* 16, 295–308.

Paul, G. 1988. *Predatory Dinosaurs of the World.* Simon & Schuster, New York.

Pawlicki, R. 1975. Studies of the fossil dinosaur bone in the scanning electron microscope. *Zeitschrift fur Mikroskopiche Anatomiche Forschun,* 89(2), 393–398.

———. 1976. Preparation of fossil bone specimens for scanning electron microscope. *Stain Technology,* 147–152.

———. 1977a. Histochemical reactions for mucopolysaccharides in the dinosaur bone: Studies on Epon- and methacrylate-embedded semithin sections as well as on isolated osteocytes and ground sections of bone. *Acta Histochemica,* 58, 75–78.

———. 1977b. Topochemical localization of lipids in dinosaur bone by means of Sudan B black. *Acta Histochemica,* 59, 40–46.

———. 1978. Morphological differentiation of the fossil dinosaur bone cells: Light, transmission electron-, and scanning electron-microscopic studies. *Acta Anatomica,* 100, 411–418.

———. 1983. Metabolic pathways of the fossil dinosaur bones. I. Vascular communication system. *Folia Histochemica et cytochemica,* 21(3–4), 253–262.

Pawlicki, R., and Bolechala, P. 1987. X-ray microanalysis of fossil dinosaur bone: Age differences in the calcium and phosphorus content of *Gallimimus bullatus* bones. *Folia Histochemica et Cytobiologica,* 25, 241–244.

Pawlicki, R., Korbel, A., and Kubiak, A. 1966. Cells, collagen fibrils and vessels in dinosaur bone. *Nature,* 211(49), 655–657.

Peabody, F. E. 1961. Annual growth zones in vertebrates (living and fossil). *Journal of Morphology,* 108, 11–62.

Petersen, H. 1930. Die Organe des skeletsystems. In *Handbuch II,* ed. W. Von Mollendorff, 521–676. Springer, Berlin.

Prum, R. O. 1999. Development and evolutionary origin of feathers. *Journal of Experimental Zoology,* 285, 291–306.

Quekett, J. 1849. On the intimate structure of bone, as comprising the skeleton in the four main classes of animals, viz., mammals, birds, reptiles and fishes, with some remarks on the great value of the knowledge of such structure in determining the affinities of minute fragments of organic remains. *Transactions of the Microscopical Society of London,* 2, 46–58.

Raath, M. A. 1969. A new coelurosaurian dinosaur from the Forest Sandstone of Rhodesia. *Arnoldia,* 4(28), 1–25.

Raath, M. A., Kitching, J. W., Shone, R. W., and Rossouw, G. J. 1990. Morphological variation in small theropods and its meaning in systematics: Evidence from *Syntarsus rhodesiensis.* In *Dinosaur Systematics: Approaches and Perspectives,* ed. K. Carpenter and P. J. Currie, 91–105. Cambridge University Press, Cambridge.

Ray, S., Botha, J., and Chinsamy, A. 2004. Bone histology and growth patterns of some non-mammalian therapsids. *Journal of Vertebrate Paleontology,* 24, 634–648.

Ray, S., and Chinsamy, A. 2004. *Diictodon feliceps* (Therapsida, Dicynodontia): Bone histology, growth and biomechanics. *Journal of Vertebrate Paleontology,* 24, 180–194.

Rayfield, E. J., Norman, D. B., Horner, C. C., Horner, J. R., Smith, P. M., Thomason, J. J., and Upchurch, P. 2001. Cranial design and function in a large theropod dinosaur. *Nature,* 409, 1033–1037.

Reid, R. E. H. 1981. Lamellar-zonal bone with zones of annuli in the pelvis of a sauropod dinosaur. *Nature,* 292, 49–51.

———. 1984a. The histology of dinosaurian bone, and its possible bearing on dinosaurian physiology. *Symposium of the Zoological Society of London,* 52, 629–663.

———. 1984b. Primary bone and dinosaurian physiology. *Geological Magazine,* 121, 589–598.

———. 1987. Bone and dinosaurian "endothermy." *Modern Geology,* 11, 133–154.

———. 1990. Zonal "growth rings" in dinosaurs. *Modern Geology,* 15, 19–48.

———. 1993. Apparent zonation and slowed late growth in a small Cretaceous theropod. *Modern Geology,* 18, 391–406.

———. 1996. Bone histology of the Cleveland-Lloyd Dinosaurs and of dinosaurs in general. I. Introduction to bone tissues. *Brigham Young University Geology Studies,* 41, 25–71.

———. 1997a. Dinosaurian physiology: The case for "intermediate" dinosaurs. In *The Complete Dinosaur,* ed. O. J. Farlow and M. K. Brett-Surman, 449–473. Indiana University Press, Bloomington.

———. 1997b. How dinosaurs grew. In *The Complete Dinosaur,* ed. J. O. Farlow and M. K. Brett-Surman, 403–413. Indiana University Press, Bloomington.

———. 1997c. A short history of dinosaur osteocytes. *Palaeontologia africana,* 34, 59–61.

Rensberger, J. M., and Watabe, M. 2000. Fine structure of bone in dinosaurs, birds, and mammals. *Nature,* 406, 619–622.

Rich, T. H. V., and Rich, P. V. 1989. Polar dinosaurs and biotas of the Early Cretaceous of Southeastern Australia. *National Geographic Research,* 5, 15–53.

Ricqlès, A. de. 1972. Recherches paléohistologiques sur les os longs de tétrapodes. II. Titanosuchiens, Dinocéphales et Dicynodontes. *Annales de Paléontologie,* 58, 1–60.

———. 1974. Evolution of endothermy: Histological evidence. *Evolutionary Theory,* 1, 51.

———. 1975. Recherches paléohistologiques sur les os longs des tétrapodes. VII. Sur la classification, la signification fonctionelle et l'histoire des tissus osseuxdes tétrapodes. Première partie: Structures. *Annales de Paléontologie,* 61, 49–129.

———. 1976. On bone histology of fossil and living reptiles, with comments on its functional and evolutionary significance. In *Morphology and Biology of Reptiles,* ed. A. d'A. Bellairs and C. B. Cox, 123–150. Academic Press, London.

———. 1980. Tissue structure of dinosaur bone: Functional significance and possible relation to dinosaur physiology. In *A Cold Look at the Warm Blooded Dinosaurs,* ed. R. D. K. Thomas and E. C. Olson, 103–139. Westview Press, Boulder, CO.

———. 1983. Cyclical growth in the long limb bones of a sauropod dinosaur. *Acta Palaeontologica Polonica,* 28, 225.

———. 1992. Paleoherpetology now: A point of view. In *Herpetology: Current Research on the Biology of Amphibians and Reptiles,* ed. K. Adler, 97–120. Proceedings of the First World Congress of Herpetology. Society for the Study of Amphibians and Reptiles, Oxford.

Ricqlès, A. de, and Buffrénil, V. de. 2001. Bone histology, heterochronies and the return of tetrapods to life in water: Where are we? In *Secondary Adaptation of Tetrapods to Life in Water,* ed. J.-M. Mazin and V. de Buffrénil, 289–310. Dr. Friedrich Pfeil, Munich.

Ricqlès, A. de, Mateus, O., Antunes, M. T., and Taquet, P. 2001. Histomorphogenesis of embryos of Upper Jurassic Theropods from Laurinhã (Portugal). *Earth and Planetary Science,* 332, 647–656.

Ricqlès, A. de, Meunier, F. J., Castanet, J., and Francillon-Vieillot, H. 1991. Comparative microstructure of bone. In *Bone.* Vol. 3, *Bone Matrix and Bone Specific Products,* ed. B. K. Hall, 1–78. CRC Press, Ann Arbor, MI.

Ricqlès, A. de, Padian, K., and Horner, J. R. 2001. The bone histology of basal birds in phylogenetic and ontogenetic perspectives. In *New Perspectives on the Origin and Early Evolution of Birds,* ed. J. Gauthier and L. F. Gall, 411–426. Proceedings of the International Symposium in Honor of John H. Ostrom. Peabody Museum of Natural History, Yale University, New Haven.

Ricqlès, A. de, Padian, K., Horner, J. R., and Francillon-Vieillot, H. 2000. Palaeohistology of the bones of pterosaurs (Reptilia: Archosauria): Anatomy, ontogeny, and biomechanical implications. *Zoological Journal of Linnean Society,* 129, 349–385.

Ricqlès, A. de, Padian, K., Horner, J. R., Lamm, E.-T., and Myhrvold, N. 2003. Osteohistology of *Confuciusornis sanctus* (Theropoda: Aves). *Journal of Vertebrate Paleontology,* 23, 373–386.

Ricqlès, A. de, Pereda Suberbiola, X., Gasparini, Z., and Olivero, E. 2001. Histology of dermal ossifications in an ankylosaurian dinosaur from the Late Cretaceous of Antarctica. Seventh International Symposium on Mesozoic Terrestrial Ecosystems. Asociación Paleontológica Argentina, Publicación especial 7, Buenos Aires, 171–174.

Rimblot-Baly, F., Ricqlès A de., and Zylberberg, L. 1995. Analyse paléohistologique d'une série de Croissance partielle chez *Lapparentosaurus madagascariensis* (Jurassic moyen): Essai sur la dynamique de Croissance d'un dinosaure Sauropode. *Annales de Paléontologie,* 81, 49–86.

Rogers, R. R., Hartman, J. H., and Krause, D. 2000. Stratigraphic analysis of Upper Cretaceous rocks in the Mahajanga Basin, northwestern Madagascar: Implications for ancient and modern faunas. *Journal of Geology,* 108, 275–301.

Rogers, R. R., and Labarbera, M. 1993. Contribution of internal bony trabeculae to the mechanical properties of the humerus of the pigeon (*Columba livia*). *Journal of Zoology, London,* 230, 433–441.

Rothschild, B. M., and Berman, D. 1991. Fusion of the caudal vertebrae in late Jurassic sauropods. *Journal of Vertebrate Paleontology,* 11, 29–36.

Rothschild, B. M., and Martin, L. D. 1993. Paleopathology: Disease in the fossil record. CRC Press, Boca Raton, LA.

Rothschild, B. M., and Tanke, D. 1991. Paleopathology of vertebrates: Insights to lifestyles and health in the geological record. *Geoscience Canada,* 19, 73–82.

Rothschild, B. M., Tanke, D., and Carpenter, K. 1997. Tyrannosaurs suffered from gout. *Nature,* 387, 357.

Rowe, T., Ketcham, R. A., Denison, C., Colbert, M., Xu, X., and Currie, P. J. 2001. The *Archaeoraptor* forgery. *Nature,* 410, 539–540.

Ruben, J. A. 1995. The evolution of endothermy in mammals and birds: From physiology to fossils. *Annual Review of Physiology,* 57, 69–95.

Ruben, J. A., and Bennett, A. A. 1987. The evolution of bone. *Evolution,* 41(6), 1187–1197.

Ruben, J. A., Hillenius, W. J., Geist, N. R., Leitch, A., and Horner, J. R. 1996. The metabolic status of some Late Cretaceous dinosaurs. *Science,* 273, 1204–1207.

Ruben, J. A., and Jones, T. D. 2000. Selective factors associated with the origin of fur and feathers. *American Zoology,* 40, 585–596.

Ruben, J. A., Jones, T. D., Geist, N. R., and Hillenius, W. J. 1997a. Lung structure and ventilation in theropod dinosaurs and early birds. *Science,* 278, 1267–1270.

Ruben, J. A., Leitch, A., Hillenius, W. J., Geist, N. R., and Jones, T. D. 1997b. New insights into the metabolic physiology of dinosaurs. In *The Complete Dinosaur,* ed. J. O. Farlow and M. K. Brett-Surman, 505–518. Bloomington, Indiana University Press.

Salgado, L. 2003. Considerations on the bony plates assigned to titanosaurs (Dinosauria, Sauropoda). *Ameghiniana,* 40, 441–456.

Sampson, S. D., Ryan, M. J., and Tanke, D. H. 1997. Craniofacial ontogeny in centrosaurine dinosaurs (Ornithischia: Ceratopsidae): Taxonomic and behavioral implications. *Zoological Journal of the Linnean Society,* 121, 293–337.

Sander, P. M. 2000. Longbone histology of the Tendaguru sauropods: Implications for growth and biology. *Paleobiology, 26*(3), 466–488.

Sander, P. M., and Tückmantel, C. 2003. Bone lamina thickness, bone apposition rates, and age estimates in sauropod humeri and femora. *Paläontologische Zeitchrift, 77,* 161–172.

Sasso Dal, C., and Signore, M. 1998. Exceptional soft-tissue preservation in a theropod dinosaur from Italy. *Nature, 392,* 383–387.

Schweitzer, M. H. 1997. Molecular paleontology: Rationale and techniques for the study of ancient biomolecules. In *The Complete Dinosaur,* ed. J. O. Farlow and M. K. Brett-Surman, 136–149. Indiana University Press, Bloomington.

Schweitzer, M. H., and Cano, R. J. 1994. Will the dinosaurs rise again? In *Dinofest,* ed. G. D. Rosenberg and D. Wolberg, 309–326. Special Publication No. 7. Paleontological Society, University of Tennessee, Knoxville.

Schweitzer, M. H., Watt, J. A., Avci, R., Knapp, L., and Chiappe, L. M. 1999. Beta-keratin specific immunological reactivity in feather-like structures of the Cretaceous alvarezsaurid *Shuvuuia deserti. Journal of Experimental Zoology, 285,* 146–157.

Seeley, H. G. 1887. On the classification of the Dinosauria. *Reports of the British Association for the Advancement of Science,* 1887, 698–699.

Seitz, A. 1907. Vergleichende Studien über den mikroskopischen Knochenbau fossiler und rezenter Reptilien und dessen Bedeutung für das Wachstum und Umbildung des Knochengewebes in allgemeinen. *Nova Acta Abhandlungen der Kaiserlichen Leopold-Carolignischen deutschen Akademie der Naturforscher,* 87, 230–270.

Sereno, P. C. 1999. The evolution of dinosaurs. *Science,* 284, 2137–2147.

Sereno, P. C., and Rao, C. 1992. Early Evolution of avian flight and perching: New evidence from the Lower Cretaceous of China. *Science,* 255, 845–848.

Sheldon, A. 1997. Ecological implications of mosasaur bone microstructure. In *Ancient Marine Reptiles,* ed. J. M. Callaway and E. L. Nicholls, 333–354. Academic Press, New York.

Smith, K. G., and Bradley, D. A. 1954. Radioactive fossil bones in Teton Country, Wyoming. *Geological Survey of Wyoming, Report of Investigations,* No. 4, 1–12.

Smith, M. M., and Hall, B. K. 1991. Development and evolutionary origins of vertebrate skeletogenic and odontogenic tissues. *Biological Reviews, 65,* 277–373.

Starck, J. M., and Chinsamy, A. 2002. Bone microstructure and developmental plasticity in birds and other dinosaurs. *Journal of Morphology,* 254, 232–246.

Swinton, W. E. 1934. *The Dinosaurs.* Thomas Murby, London.

Thomason, J. J. 1995. To what extent can the mechanical environment of a bone be inferred from its internal architecture? In *Functional Morphology in Vertebrate Paleontology,* ed. J. J. Thomason, 249–263. Cambridge University Press, New York.

Trueman, C. N., and Martill, D. M. 2002. The long-term survival of bone: The role of bioerosion. *Archaeometry,* 44, 371–382.

Trueman, C. N., and Tuross, N. 2003. Trace elements in recent and fossil bone apatite. *Reviews in Mineralogy and Geochemistry,* 48, 489–521.

Tumarkin-Deratzian, A. 2003. Bone Surface Texture as Ontogenetic Indicators in Extant and Fossil Archosaurs: Macroscopic and Histological Evaluations. PhD thesis. University of Pennsylvania, Philadelphia.

Varricchio, D. J. 1993. Bone microstructure of the Upper Cretaceous theropod dinosaur *Troodon formosus*. *Journal of Vertebrate Paleontology,* 13, 99–104.

———. 1997. Growth and embryology. In *Encyclopedia of Dinosaurs,* ed. P. J. Currie and K. Padian, 282–288. Academic Press, London.

———. 1997. Troodontidae. In *Encyclopedia of Dinosaurs,* ed. P. J. Currie and K. Padian, 749–759. Academic Press, London.

Varricchio, D. J., Jackson, F., Borkowski, J. J., and Horner, J. R. 1997. Nest and egg clutches of the dinosaur *Troodon formosus* and the evolution of avian reproductive traits. *Nature,* 385, 247–250.

Wall, W. P. 1983. The correlation between limb-bone density and aquatic habits in recent mammals. *Journal of Paleontology,* 57(2), 197–207.

Wedel, M. J. 2003. Vertebral pneumacity, air sacs, and the physiology of sauropod dinosaurs. *Paleobiology,* 29, 243–255.

Weidenreich, F. 1930. Das knochengewebe. In *Handbuch der Mikorskopeschen,* ed. W. von Mollendorff, 391–520. Springer, Berlin.

Weiner, S., and Traub, W. 1992. Bone structure: From angstroms to microns. *FASEB Journal,* 6, 879–885.

Winkler, D. A. 1994. Aspects of growth in the early cretaceous proctor lake ornithopod. *Journal of Vertebrate Paleontology (Abstracts),* 14, 53A.

Witmer, L. M. 1995a. The extant phylogenetic bracket and the importance of reconstructing soft tissues in fossils. In *Functional Morphology in Vertebrate Paleontology,* ed. J. J. Thomason, chap. 2. Cambridge University Press, Cambridge.

———. 1995b. Homology of facial structures in extant archosaurs (birds and crocodilians), with special reference to paranasal pneumaticity and nasal conchae. *Journal of Morphology,* 225, 269–327.

———. 2001. Nostril position in dinosaurs and other vertebrates, and its significance for nasal function. *Science,* 293, 850–853.

———. 2002. The debate on avian ancestry: Phylogeny, function and fossils. In *Mesozoic Birds: Above the Heads of Dinosaurs,* ed. L. M. Chiappe and L. M. Witmer, 3–30. University of California Press, Berkeley.

Witmer, L. M., Chatterjee, S., Franzosa, J., and Rowe, T. 2003. Neuroanatomy of flying reptiles and implications for flight, posture and behaviour. *Nature,* 425, 950–953.

Woo, S. L. Y., Kuei, S. C., Amiel, D., Gomez, M. A., Hayes, W. C., White, F. C., and Akeson, W. H. 1981. The effect of prolonged physical training on the properties of long bone: A study of Wolff's law. *Journal of Bone and Joint Surgery,* 63, 780–787.

Woodward, S. R., Weyand, N. J., and Bunnell, M. 1994. DNA sequence from Cretaceous period bone fragments. *Science,* 266, 1229–1232.

Xu, X., Zhou, Z., Wang, X., Kuang, X., Zhang, F., and Du, X. 2003. Four-winged dinosaurs from China. *Nature,* 421, 335–340.

Yao, J., and Zhang, Y. 1997. Histological study on the Late Cretaceous Dinosaur's bones, and comparison with modern reptilian and avian bones. *VertebrataPal Asiatica,* 35(3), 178–181.

Zhang, F., Hou, L., and Ouyang, L. 1998. Osteological microstructure of *Confuciusornis:* Preliminary report. *Vertebrata PalAsiatica,* 36(2), 133–136.

Zhang, F., Xu, X., Lu, J., and Ouyang, L. 1999. Some microstructure difference among *Confuciusornis,* alligator and a small theropod dinosaur, and its implications. *Palaeoworld,* 11, 308.

Zhou, Z., and Zhang, F. 2003. *Jeholornis* compared to *Archaeopteryx,* with a new understanding of the earliest avian evolution. *Naturwissenschaften,* 90, 220–225.

Index

Figure page locators and plate numbers are in boldface.